Mathematische Optimierung und Wirtschaftsmathematik | Mathematical Optimization and Economathematics

Reihe herausgegeben von

R. Werner, Augsburg, Deutschland

T. Harks, Augsburg, Deutschland

V. Shikhman, Chemnitz, Deutschland

In der Reihe werden Arbeiten zu aktuellen Themen der mathematischen Optimierung und der Wirtschaftsmathematik publiziert. Hierbei werden sowohl Themen aus Grundlagen, Theorie und Anwendung der Wirtschafts-, Finanz- und Versicherungsmathematik als auch der Optimierung und des Operations Research behandelt. Die Publikationen sollen insbesondere neue Impulse für weitergehende Forschungsfragen liefern, oder auch zu offenen Problemstellungen aus der Anwendung Lösungsansätze anbieten. Die Reihe leistet damit einen Beitrag der Bündelung der Forschung der Optimierung und der Wirtschaftsmathematik und der sich ergebenden Perspektiven.

Weitere Bände in der Reihe http://www.springer.com/series/15822

Emilia Graß

An Accelerated Solution Method for Two-Stage Stochastic Models in Disaster Management

Springer Spektrum

Emilia Graß
Institute for Operations Research
and Information Systems
Hamburg University of Technology
Hamburg, Germany

Dissertation, Hamburg University of Technology, 2018

ISSN 2523-7926 ISSN 2523-7934 (electronic)
Mathematische Optimierung und Wirtschaftsmathematik | Mathematical Optimization and Economathematics
ISBN 978-3-658-24080-6 ISBN 978-3-658-24081-3 (eBook)
https://doi.org/10.1007/978-3-658-24081-3

Library of Congress Control Number: 2018960286

Springer Spektrum

This Springer Spektrum imprint is published by the registered company Springer Fachmedien Wiesbaden GmbH part of Springer Nature
The registered company address is: Abraham-Lincoln-Str. 46, 65189 Wiesbaden, Germany

Summary

Natural disasters cause enormous damages to humans, the economy and the environment every year. Aid agencies try to improve the effectiveness of relief operations by implementing appropriate preparation strategies before a disaster occurs. Pre-positioning of relief items at selected locations, e.g. public buildings like gyms or hospitals, is one of the most important tasks in the preparation phase. Since such storage and location decisions have to be made under a high degree of uncertainty, two-stage stochastic programs are often proposed in the literature to support decision makers. These models contain two types of decisions: Decisions made before the realization of an uncertain event like a natural disaster are referred to as first-stage decisions. In the aftermath of a disaster, where additional relevant information becomes available, second-stage decisions can be made. For instance, decisions regarding purchase and storage of relief items before the occurrence of a disaster belong to the first stage. At this stage, there is uncertainty regarding the location, time and magnitude of a disaster. As soon as the corresponding information on, e.g., actual demand and damage level is known, relief items can be reordered and pre-positioned items can be delivered to satisfy demand at affected locations.

However, two-stage stochastic problems are often very large in practical applications, preventing commercial solvers from providing the corresponding solution within an acceptable time. In this doctoral thesis, an efficient solution method is introduced which overcomes several disadvantages of heuristics as well as exact algorithms and leads to fast near-optimal solutions of large-scale two-stage stochastic programs. The main idea is to accelerate a well-known solution method for this kind of problems, namely the L-Shaped method, by exploiting the specific structure of the second-stage sub-problems. In particular, a specialized interior-point method is proposed to solve the sub-problems fast. The advantage of the accelerated L-shaped method is its applicability to a whole class of two-stage stochastic models often proposed in the humanitarian literature.

Humanitarian organizations can use the accelerated L-shaped method to improve their agility and reactivity, especially when disaster forecasts are available. Indeed, floods and hurricanes can be predicted three to five days in advance. Within this time frame, solving realistic problems rapidly is essential for aid agencies to take appropriate preparation actions. In order to demonstrate the computational benefits of the newly proposed solution approach problems of different problem sizes are solved. A realistic large-scale case study is designed for the hurricane prone Gulf- and Atlantic coast. Moreover, the three-day forecast for hurricane Katrina in 2005 is used to define a second realistic case study.

The accelerated solution method enables aid agencies to apply two-stage stochastic models in preparation for short-term forecasts, alleviating human suffering and potentially preventing casualties.

Contents

List of Abbreviations XIII

List of Symbols XV

1 Introduction 1

2 Two-Stage Stochastic Programs for Pre-Positioning Problems in Disaster Management 3
2.1 Disaster Management . . . 3
2.1.1 Introduction . . . 3
2.1.2 Challenges . . . 9
2.1.3 Scenario Definition in Disaster Management . . . 10
2.2 Quantitative Models in Disaster Management: A Literature Review . . . 11
2.2.1 Two-Stage Stochastic Programs . . . 11
2.2.2 Pre-Positioning of Relief Items . . . 28
2.3 The Rawls and Turnquist [2010] Model . . . 30
2.3.1 Problem Description and Mathematical Formulation . . . 30
2.3.2 Extensions . . . 35

3 Solution Algorithms in Disaster Management 37
3.1 Solution Methods in Disaster Management: A Literature Review . . . 38
3.1.1 Exact Methods . . . 39
3.1.2 Heuristics . . . 40
3.2 Two-Stage Stochastic Programming . . . 45
3.2.1 Introduction . . . 45
3.2.2 The L-Shaped Method . . . 49
3.3 The Accelerated L-Shaped Method . . . 55
3.3.1 The Basic Idea . . . 55
3.3.2 Assumptions . . . 58
3.3.3 Specialized Primal-Dual Interior-Point Method . . . 60

4 Numerical Experiments 71
4.1 Realistic Large-Scale Case Study . . . 71
4.1.1 Data . . . 73
4.1.2 Technical Specifications . . . 82
4.1.3 Computational Results . . . 84
4.2 Case Study Based on a Hurricane Forecast . . . 97
4.2.1 Data . . . 98

4.2.2 Computational Results 101
4.3 Outlook 104

5 Conclusion **107**

Bibliography **111**

A Appendix **125**
A.1 The Recourse Function: An Example 125
A.2 Newton's Method for Systems of Non-Linear Equations 126
A.3 Interior-Point Method: Proof of Convergence 127
A.4 Matlab Code: L-Shaped Method with Multi-Optimality Cuts 130
A.5 Matlab Code: SIMP 136
A.6 Gurobi Log Files 140
A.6.1 Small-Scale Case Study 140
A.6.2 Medium-Scale Case Study 143
A.6.3 Large-Scale Case Study 146
A.6.4 Katrina Case Study 152

List of Figures

2.1 Percentage of occurrence of natural disasters from 1995 to 2015 [CRED/ UNISDR, 2015] 4
2.2 Consequences of natural disasters from 2000 to 2016 (based on EM-DAT [2017]) 5
2.3 Phases in disaster management (based on Kovács and Spens [2007]) 5
2.4 United Nations Humanitarian Response Depots [WFP, 2017] 6
2.5 Partner hospitals of Direct Relief [2017] 7
2.6 Decisions in a two-stage stochastic problem for a hurricane example (based on Schade [2012, p. 46]). 12
2.7 Network containing nodes and arcs using the example of the Gulf- and Atlantic coastal plains [Rawls and Turnquist, 2010] 31

3.1 Example of a piece-wise linear expected recourse function $E_\xi\left[Q(\chi, \xi(s))\right]$. 47
3.2 The L-shaped method (based on Birge and Louveaux [2011, Ch. 5]) 51
3.3 Optimality cuts (OC) approximating the expected recourse function 52
3.4 Sub-gradients of a non-smooth function (based on Boyd and Vandenberghe [2004, p. 69]) 53
3.5 The accelerated L-shaped method [Grass et al., 2018] 57
3.6 The central path $\mathcal{C}$ with γ_μ, depending on different values of μ, i.e. $\mu = \infty$, $\mu = 1$ and $\mu = 0.1$ (based on Kwon [2013, p. 223]) 62
3.7 A primal-dual path-following interior-point method 64

4.1 Counties hit by hurricanes of different categories from 1900-2014 (based on IBTrACS) 73
4.2 Network containing 90 nodes on the Gulf and Atlantic coast 74
4.3 Landfall nodes of major (red circles) and minor (yellow circles) hurricanes in Rawls and Turnquist [2010] 77
4.4 Sparsity pattern of the recourse matrix W in (3.5) where nz denotes the number of non-zero entries 80
4.5 Enlarged extracts of the recourse matrix W showing the first upper part (a) and last lower part (b) in Figure 4.4 80
4.6 Sparsity pattern of the Schur complement $WD^{-1}W^T$ for the large-scale case study 81
4.7 Scheme of the standard and accelerated L-shaped method 83
4.8 MATLAB's profiler showing the most time-consuming functions of the standard L-shaped method for the small-scale case study 86
4.9 Facility locations as suggested by Rawls and Turnquist [2010] (top) and for the small-scale case study (bottom) 87

4.10 MATLAB's profiler showing the most time-consuming functions of the standard L-shaped method for the medium-scale case study 89
4.11 Upper bound (UB) and lower bound (LB) for the medium-scale case study 90
4.12 MATLAB's profiler showing the most time-consuming operations of the standard L-shaped method for the large-scale case study 92
4.13 MATLAB's profiler showing the most time-consuming operations of the accelerated L-shaped method for the large-scale case study 92
4.14 MATLAB's profiler showing the most time-consuming operations of SIPM for the large-scale case study 93
4.15 Condition number of saddle point matrix S, relative residuals and optimality gap 94
4.16 Facility locations as suggested by Rawls and Turnquist [2010] (top) and for the medium-scale case study (center) and the large-scale case study (bottom) 96
4.17 Predicted path of hurricane Katrina, 26 August 2005 at 5pm (three days before landfall [NHC, 2005]) 98
4.18 Potentially vulnerable nodes based on the predicted path of hurricane Katrina 99
4.19 Predicted category and probability of hurricane Katrina, 26 August 2005 at 5pm (three days before landfall [NHC, 2005]) 100
4.20 Suggested facility locations for hurricane Katrina three days before landfall 102
4.21 Actual path of hurricane Katrina (red dashed line), counties affected to varying extents and suggested facility locations 103

A.1 Piece-wise linear and convex recourse function $Q(\chi)$ for the given second-stage problem 125

List of Tables

2.1 Publications classified according to disaster phase and type 14
2.1 Publications classified according to disaster phase and type (continued) . . 15
2.1 Publications classified according to disaster phase and type (continued) . . 16
2.2 Objective functions and decisions for first and second stage in the pre-disaster phase 17
2.2 Objective functions and decisions for first and second stage in the pre-disaster phase (continued) 18
2.2 Objective functions and decisions for first and second stage in the pre-disaster phase (continued) 19
2.3 Uncertainties, problematic assumptions and special features of pre-disaster models 21
2.3 Uncertainties, problematic assumptions and special features of pre-disaster models (continued) 22
2.3 Uncertainties, problematic assumptions and special features of pre-disaster models (continued) 23
2.4 Objective functions and decisions for first and second stage of post-disaster phase 27
2.5 Uncertainties, problematic assumptions and special features of post-disaster models 27
2.6 Scenario-free approaches: objectives, uncertainties, approaches capturing uncertainties and assumptions 29

3.1 Solution methods 38
3.1 Solution methods (continued) 39
3.2 Types of heuristics 41
3.3 Model type of the second stage 59

4.1 Hurricane categories [NHC, 2017b] 72
4.2 Opening costs and storage capacities [Rawls and Turnquist, 2010] 74
4.3 Purchase price, weight and space requirement of each relief item (based on IFRC [2016] and SCEMD [2007]) 75
4.4 Percentage of population to be provided with relief items depending on the damage potential of hurricanes [NOAA and AOML, 2016] 78
4.5 Required relief items per person for two days 78
4.6 Data sources of the case study proposed by Rawls and Turnquist [2010] and the newly proposed case study 79
4.7 Problem size of a small-, medium- and large-scale case study 82

4.8 Performance overview of different solution approaches for the small-scale case study . . . 85
4.9 Performance overview of different solution approaches for the medium-scale case study . . . 88
4.10 Performance overview of different solution approaches for the large-scale case study . . . 91
4.11 Performance overview of different solution approaches for the case study based on the short-term forecast . . . 101

List of Abbreviations

AL	Alabama
AOML	Atlantic Oceanographic and Meteorological Laboratory
CARE	Cooperative for Assistance and Relief Everywhere
CB	Census Bureau
CPLEX	IBM ILOG CPLEX Optimization Studio
CPU	Central Processing Unit
CRED	Centre for Research on the Epidemiology of Disasters
cVaR	conditional Value-at-Risk
EM-DAT	Emergency Events Database
FAK	First-Aid Kit
FEMA	Federal Emergency Management Agency
FL	Florida
GAMS	General Algebraic Modeling System
GDP	Gross Domestic Product
HLT	Hurricane Liaison Team
IBTrACS	International Best Track Archive for Climate Stewardship
IFRC	International Federation of Red Cross and Red Crescent Societies
KKT	Karush–Kuhn–Tucker
LA	Louisiana
LB	Lower Bound
LLSM	Lagrangian L-Shaped Method
LP	Linear Program
MILP	Mixed Integer Linear Program
MIP	Mixed Integer Progam
MRE	Meal Ready-to-Eat
MS	Mississippi
MSF	Médecins Sans Frontières (Doctors without Borders)
NHC	National Hurricane Center
NHCFP	National Hurricane Center Forecast Process
NWS	National Weather Service
OC	Optimality Cuts
OV	Objective Value
PDSA	Percentage of Demand Scenarios solution Approach
PH	Progressive Hedging
PSO	Particle Swarm Optimization
RelOptGap	Relative Optimality Gap
rhs	right hand side

SCEMD	South Carolina Emergency Management Division
SIPM	Specialized Interior-Point Method
SRT	Standard Relief Time
SUMA	Humanitarian Supply Management System
UB	Upper Bound
UNHRD	United Nations Humanitarian Response Depot
UNICEF	United Nations Children's Fund
UNISDR	United Nations Office for Disaster Risk Reduction
URI GSO	University of Rhode Island and Graduate School of Oceanography
WFO	Weather Forecast Office
WFP	World Food Programme

List of Symbols

	Mathematical Operators
∇	Nabla operator
∂	sub-differential
$\triangle$	Newton search direction
	Greek Letters
α	predefined probability for chance constraints
$\alpha^{\mathbf{k}}$	step length in interior-point iteration $\mathbf{k}$
γ	vector of second-stage decision variables for the primal problem
Γ	diagonal matrix with vector γ on its diagonal
η	step length damping parameter
ϑ	probability weighted second-stage objective value for given first-stage solutions
θ	approximation for probability weighted second-stage objective value
ι	L-shaped iteration counter
λ	risk parameter
μ	barrier parameter
ξ	vector containing uncertain data
π	vector of Lagrange/Simplex multipliers and dual decision variables for the second-stage problem
ρ_i^{ks}	proportion of relief item $k \in \mathrm{K}$ at node $i \in \mathrm{I}$ in scenario $s \in \mathrm{S}$ which is still usable
σ	centering parameter
τ	duality measure
χ	vector of first-stage decision variables
ψ^s	1 if scenario $s \in \mathrm{S}$ is part of the reliable set; 0 otherwise
Ω	set of random events
	Latin and Mixed Letters
a	first-stage cost rates
A	(1,1)-block matrix of $WD^{-1}W^T$
b^k	unit space requirement for relief item $k \in \mathrm{K}$
B	(2,2)-block matrix of $WD^{-1}W^T$
$c_{(ij)}^k$	costs for transporting one unit of relief item $k \in \mathrm{K}$ on link $(i,j) \in \mathrm{R}$
$\mathcal{C}$	primal-dual central path
C	(1,2)-block matrix of $WD^{-1}W^T$

d	first-stage right hand side
$d_{(ij)}$	distance between nodes $i \in \mathrm{I}$ and $j \in \mathrm{I}$
D	diagonal matrix
D^k	acceptable distance limit for item $k \in \mathrm{K}$
E	expectation
e	vector of all ones
f	second-stage cost rates
F_l	costs for opening facility of size $l \in \mathrm{L}$
g	sub-gradient
$g_{(i,j)}$	slack variables for link $(i, j) \in \mathrm{R}$
G	matrix of first-stage constraints
h^k	penalty cost per unit for unused relief item $k \in \mathrm{K}$
H	block matrix of recourse matrix W
I	set of demand nodes
I	identity matrix
K	set of relief items
$\mathbf{k}$	interior-point iteration counter
$\mathcal{L}$	lower triangular matrix
L	set of facility size categories
$L(\dot{})$	Lagrangian function
M_l	capacity of a facility of size $l \in \mathrm{L}$
N	set of nodes, $\mathrm{I} \subseteq \mathrm{N}$
O	large number
OV_E	objective value of the exact method
OV_H	objective value of the heuristic method
p^k	penalty cost per unit for unsatisfied demand of relief item $k \in \mathrm{K}$
P_s	probability of scenario $s \in \mathrm{S}$
q^k	acquisition cost per unit for relief item $k \in \mathrm{K}$
Q	block matrix of recourse matrix W
$Q(\chi, \xi)$	recourse function of the linear second-stage problem
r	residual
r_i^k	pre-positioned quantity of relief item $k \in \mathrm{K}$ at node $i \in \mathrm{I}$
R	set of arcs in the network
rhs	right hand side
$\mathcal{S}$	saddle point matrix
$\tilde{\mathcal{S}}$	Schur complement of matrix $WD^{-1}W^T$
S	set of possible scenarios
t_s	second-stage right hand side for scenario $s \in S$
T	technology matrix
u	block matrix of recourse matrix W
$U^s_{(ij)}$	maximal available arc capacity of link $(i,\ j) \in \mathrm{R}$ in scenario $s \in \mathrm{S}$
v_i^{ks}	demand for relief item $k \in \mathrm{K}$ at node $i \in \mathrm{I}$ in scenario $s \in \mathrm{S}$
w_i^{ks}	unsatisfied demand for relief item $k \in \mathrm{K}$ at node $i \in \mathrm{I}$ in scenario $s \in \mathrm{S}$
W	recourse matrix

$x_{(ij)}^{ks}$	quantity of relief item $k \in \mathrm{K}$ transported from $i \in \mathrm{I}$ to $j \in \mathrm{I}$ in scenario $s \in \mathrm{S}$
y_{il}	1 if facility of size $l \in \mathrm{L}$ is opened at node $i \in \mathrm{I}$; 0 otherwise
z	vector of slack variables for the dual sub-problem
z_i^{ks}	unused quantity of relief item $k \in \mathrm{K}$ at node $i \in \mathrm{I}$ in scenario $s \in \mathrm{S}$
Z	diagonal matrix with vector z on its diagonal

1 Introduction

Recent series of natural disasters have led to countless fatalities, ecological as well as infrastructural devastation and economic damage. Prominent examples are the three major earthquakes in Mexico with 8.1, 7.1 and 6.1 magnitude on the Richter scale all in September 2017 or the three American hurricanes *Harvey*, *Irma* and *Maria* of the highest category occurring within a very short time in August and September 2017. The American 2017 hurricane season has been one of the most devastating and expensive seasons so far, resulting in about US$ 290 billion of economic damage [AccuWeather, 2017]. Due to climate change the number and severity of natural disasters have increased significantly and will even continue to rise [Coppola, 2011, p. 25]. Although it is impossible to prevent catastrophic events, their devastating consequences can be reduced by implementing appropriate preparation strategies like initiating evacuations and providing temporary shelters before disasters occur. Disaster management is concerned with developing strategies for mitigating negative impacts before, during and after catastrophic events. Pre-positioning of relief items, e.g. water and medication, at selected locations is regarded as the most important preparation measure. Often facility and pre-positioning decisions are taken without prior planning, resulting in a potentially delayed and insufficient response [Balcik and Beamon, 2008]. Due to such ad-hoc decisions high costs may arise as well as uncoordinated and thereby insufficient or redundant stocks. In contrast, determining the locations of relief facilities and the amount of aid items to be stored before the occurrence of an emergency can accelerate the adequate supply to those affected and hence decrease the number of casualties. However, decisions concerning the location and amount of aid supplies have to be made under a high degree of uncertainty. At the time of preparation decisions, the disaster's exact location, time and magnitude are unknown. In order to support decision makers in this challenging task, a variety of quantitative optimization models are proposed in the humanitarian literature. In particular, two-stage stochastic programming is one of the most popular modeling approaches in stochastic programming [Shapiro and Philpott, 2007] and especially in disaster management [Falasca and Zobel, 2011]. Two-stage stochastic models contain two types of decisions. First-stage decisions have to be made *before* and second-stage decisions *after* the occurrence of a random event. On the first-stage, i.e. where relevant data is not yet available, decisions are made under uncertainty. As soon as uncertainty has unfolded, so-called *recourse* decisions on the second stage can be carried out, depending on the impact of the disaster and given first-stage decisions. For instance, on the first stage relief items can be purchased in preparation for a particular hurricane, representing the uncertain event. After the hurricane's actual path and intensity, i.e. damage on infrastructure and demand become known, items pre-positioned on the first stage have to be delivered to affected regions and additional relief items may have to be purchased to satisfy actual demand.

E. Graß, *An Accelerated Solution Method for Two-Stage Stochastic Models in Disaster Management*, Mathematische Optimierung und Wirtschaftsmathematik | Mathematical Optimization and Economathematics,
https://doi.org/10.1007/978-3-658-24081-3_1

Although two-stage stochastic models are very useful in practice they are often very large for real-world applications and therefore commercial solvers may fail to provide the corresponding solution within an acceptable time frame. The main objective of this doctoral thesis is to develop an accelerated solution method which overcomes several disadvantages of heuristics and exact algorithms and can lead to fast solutions of large-scale two-stage stochastic programs resulting in humanitarian settings. Relief organizations can use these fast solutions to improve their agility and reactivity, especially when disaster forecasts are available. Indeed, floods and hurricanes can be predicted three to five days in advance [Environment Agency, 2017] [NOAA and AOML, 2016]. Solving realistic problems rapidly is therefore essential for employing appropriate measures in preparation for imminent threats.

The thesis is organized as follows. The next chapter is mainly concerned with two-stage stochastic programs for pre-positioning problems in disaster management. After giving a general introduction to disaster management (Section 2.1), quantitative models proposed in the humanitarian literature are reviewed (Section 2.2). As pre-positioning of relief items plays a crucial part in the preparation phase, special emphasis is placed on the advantages of storing aid supplies in advance. The two-stage stochastic model proposed by Rawls and Turnquist [2010], which is presented in Section 2.3, is one of the first and most cited in the pre-positioning literature. For this reason, the Rawls and Turnquist [2010] model forms the basis for the theoretical development of an accelerated solution method and for numerical experiments in the following.

The success of a relief operation in emergency situations depends strongly on the availability of fast solutions for realistic large-scale problems. Therefore, developing an accelerated solution algorithm is the focus of Chapter 3. At the beginning, existing solution methods for two-stage stochastic models in disaster management are reviewed and critically evaluated (Section 3.1). Based on the theoretical principles of two-stage stochastic programming provided in Section 3.2, the *accelerated L-shaped method* is developed in Section 3.3. The L-shaped method is a well-known algorithm for solving two-stage stochastic programs. In order to improve this standard approach, a *specialized interior-point method* (SIPM) is proposed. The main idea of SIPM is to exploit the specific structure of the second-stage problem, resulting in an accelerated version of the standard L-shaped method.

In order to verify the performance of the theoretically derived accelerated L-shaped algorithm, numerical tests for different problem sizes are carried out in Chapter 4. In particular, a realistic large-scale case study based on the one given in Rawls and Turnquist [2010] is designed in Section 4.1. A second case study is presented in Section 4.2, where the three-day forecast for hurricane Katrina in 2005 is used. Especially with this example, the usefulness of the approach in real emergencies can be illustrated. An outlook on possible future developments if problem sizes increase even further is given at the end of this chapter (Section 4.3).

The thesis concludes in Chapter 5 with a discussion on the advantages of the newly proposed solution method and with recommendations for future research.

2 Two-Stage Stochastic Programs for Pre-Positioning Problems in Disaster Management

Disasters pose a large threat to the environment and human society. Disaster management is concerned with the development of all activities and measures to prevent and alleviate the dramatic consequences of disasters. The main priorities and strategies of disaster management are introduced in Section 2.1.1. Unfortunately, a variety of problems arise in the implementation of relief efforts, complicating the work of aid organizations significantly. Based on the discussion on major challenges in Section 2.1.2, the uncertain nature of disasters represents the most crucial issue for relief managers. In particular, decisions have to be made under a high degree of uncertainty. In this regard, two-stage stochastic models play a key role in supporting decision makers. The main ingredient of such models is the definition of appropriate scenarios, as stated in Section 2.1.3.

Section 2.2.1 reviews two-stage stochastic models in the state-of-the-art literature for the pre- and post-disaster phase. As indicated by this survey, pre-positioning of relief items belongs to the most addressed preparation strategies within disaster management. The advantages of pursuing this strategy are highlighted in Section 2.2.2.

The model proposed by Rawls and Turnquist [2010] is a two-stage stochastic program; this is one of the most cited articles in the pre-positioning literature. Therefore, it is used as the reference model in this regard. The corresponding mathematical formulation can be found in Section 2.3.1 whereas its extensions proposed in the literature are described in 2.3.2.

2.1 Disaster Management

2.1.1 Introduction

According to the Centre for Research on the Epidemiology of Disasters (CRED) a disaster is: "*a situation or event which overwhelms local capacity, necessitating a request to a national or international level for external assistance; an unforeseen and often sudden event that causes great damage, destruction and human suffering*" [Guha-Sapir et al., 2012]. In general, disasters can have many different causes. Based on the classification of Coppola [2011, p. 38] disasters can be divided into *natural*, *technological* and *intentional* disasters. Natural disasters may have different causes like geological processes leading to earthquakes and volcanic eruption or meteorological factors triggering storms and

E. Graß, *An Accelerated Solution Method for Two-Stage Stochastic Models in Disaster Management*, Mathematische Optimierung und Wirtschaftsmathematik | Mathematical Optimization and Economathematics,
https://doi.org/10.1007/978-3-658-24081-3_2

hurricanes [Alexander, 2002, p. 3]. In contrast, technological or man-made disasters have their origins in technological innovations. These may include chemical accidents, leakage of oil, or nuclear reactor accidents [Ortuño et al., 2013]. Terrorist attacks belong to the category of intentional hazards, where humans use weapons or violence with the intent to harm people and societies [Coppola, 2011, p. 118]. Since natural disasters are regarded to be most common [Coppola, 2011, p. 38], only this type of hazard is considered in the following.

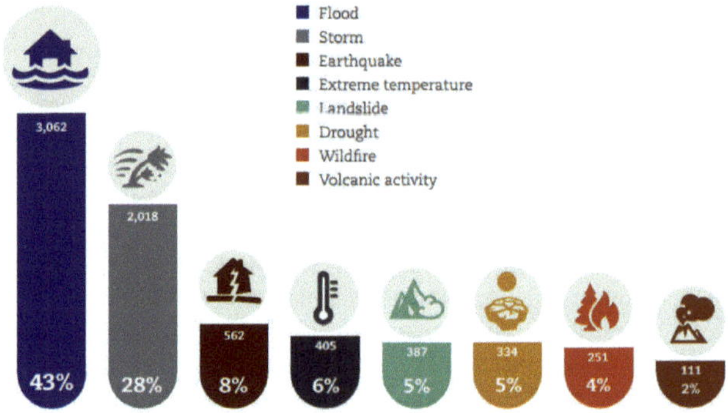

Figure 2.1: Percentage of occurrence of natural disasters from 1995 to 2015 [CRED/ UNISDR, 2015]

As shown in Figure 2.1, among the natural disasters floods and storms occurred most often in the last 20 years. The high proportion of floods, namely 43%, may be explained by the fact that floods occur either solely or as a "by-product" of storms [NHC, 2017a]. Indeed, storms and hurricanes are most often accompanied by torrential downpours and flooding, causing more casualties than the actual storm. Although earthquakes account for only 8% of all disaster types, their impact on society is devastating.

In general, humans and economy are suffering severely from the negative impacts of all kinds of natural disasters every year. Figure 2.2 highlights the dramatic influence of natural disasters on human society in the period between 2000 and 2016. For instance, over one million people were killed and almost four million were affected during this time. The economic damage, defined as "*losses as a percentage of Gross Domestic Product (GDP)*" [CRED/ UNISDR, 2015], reached 5% of GDP for low-income countries and amounted to nearly US$ 2 trillion for all countries in the last 16 years. As indicated by Figure 2.2, the number of killed persons and the amount of economic damage is not necessarily correlated. One of the reasons is the level of development of the respective country. For instance, the flood in Pakistan and the earthquake in Haiti, both in 2010, claimed most lives in the past. The corresponding economic damage of around US$ 132 billion was, however, relatively low. In contrast, Japan's nuclear disaster caused by an earthquake and a tsunami in 2011 represents the most expensive disaster of all times. Despite this economic impact the number of deaths was below other major disasters in the past.

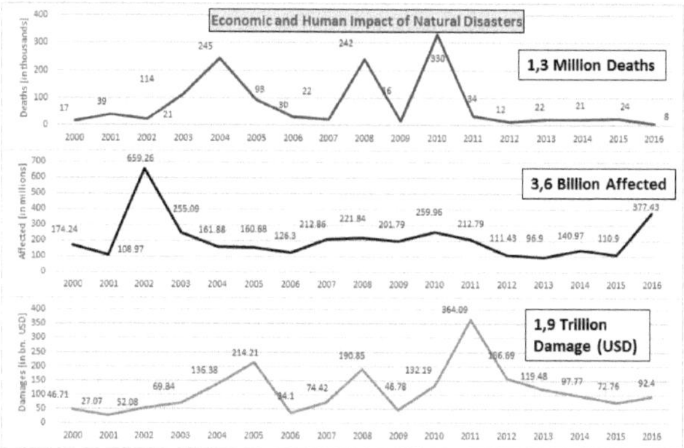

Figure 2.2: Consequences of natural disasters from 2000 to 2016 (based on EM-DAT [2017])

According to Coppola [2011, p. 25] the number of disaster occurrences is even rising, having an ever greater impact on the society. Therefore, strategies for mitigating the immense consequences of natural disasters have to be developed. Disaster management makes an important contribution in this regard by employing appropriate measures corresponding to three phases: preparation, immediate response and reconstruction phase.

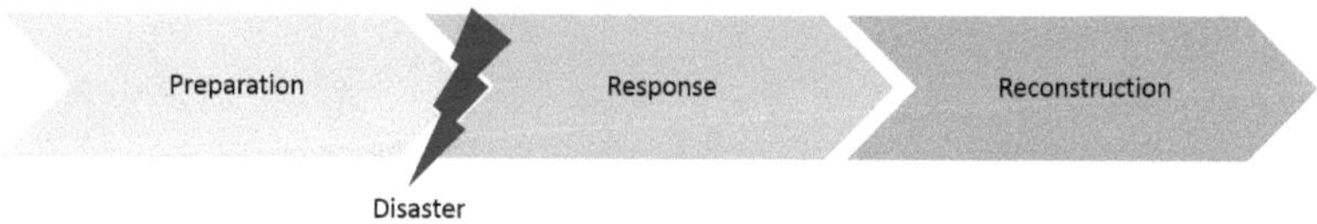

Figure 2.3: Phases in disaster management (based on Kovács and Spens [2007])

As one can see in Figure 2.3 the preparation phase precedes a disaster whereas the response and reconstruction phases take effect after the disaster occurrence. In the following, strategies for each phase are described in more detail.

- Preparation Phase
  The preparation phase contains all activities necessary for resisting and combating hazards and damage of potential disasters [Coppola, 2011, p. 251]. Such activities are especially useful for disaster-prone areas like the Gulf- and Atlantic coastal plains which are hit by storms and hurricanes every year. Central Italy and Japan are examples for earthquake-prone countries whereas India is faced with recurrent floods in rainy seasons.

Constructing hazard-resistant buildings or strengthening existing bridges or roads can reduce the vulnerability significantly [Coppola, 2011, pp. 214].

Generally, trainings for locals and humanitarian teams for correct responses in emergency situations are offered in the preparation phase [Van Wassenhove, 2006]. These include, for example, emergency exercises and evacuation trainings. Before evacuation can be exercised or executed, appropriate evacuation plans have to be prepared. Based on these preliminary plans people at risk can be either evacuated when a concrete disaster is predicted or after a sudden-onset disaster.

Providing first aid supplies to the affected population as fast as possible is crucial in the aftermath of a disaster. In order to achieve a rapid response in emergency situations essential relief items should be available immediately. This can be achieved by pre-positioning these items at selected locations in advance. Based on the report of Adinolfi et al. [2005] such pre-positioning strategies are pursued by 60% of relief organizations surveyed. For instance, the aid organization UNICEF ensures that their warehouses are sufficiently stocked at any time [Dignan, 2005]. Regions hit by a disaster can be reached from the six United Nations Humanitarian Response Depots (UNHRD), shown in Figure 2.4, within 24-48 hours. The World Food Programme (WFP) located these depots worldwide taking convenient connections to airports, ports and main roads into account [WFP, 2017].

Figure 2.4: United Nations Humanitarian Response Depots [WFP, 2017]

Although pre-positioning of relief items can accelerate responsiveness significantly, establishing and maintaining warehouses can be too expensive especially for smaller aid agencies [Balcik and Beamon, 2008]. Therefore, contracts and agreements with suppliers are often signed, avoiding the need for own inventories. In an emergency situation, aid supplies can then be ordered and stored temporarily at available public buildings like gyms, city halls or hospitals.

The American aid organization *Direct Relief* has set up a Hurricane Preparedness Program with partner hospitals in the U.S. Gulf Coast, Central America and Caribbean [Direct Relief, 2017]. The orange icons in Figure 2.5 represent its 50 partner health-care institutions mainly in the Gulf- and Atlantic coast areas.

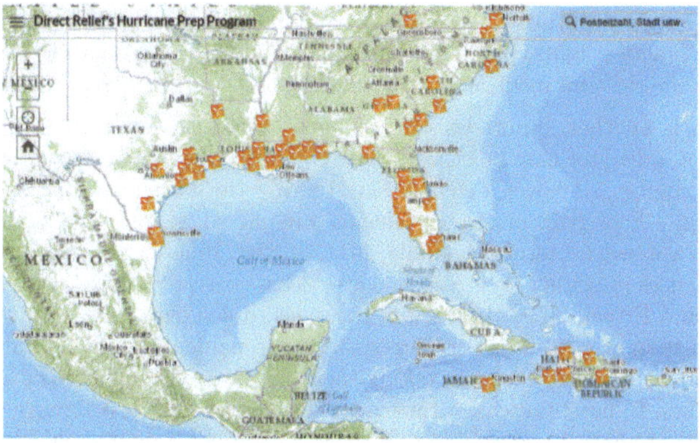

Figure 2.5: Partner hospitals of Direct Relief [2017]

Every year Direct Relief supplies its partners with medical equipment like medicine and syringes for the upcoming hurricane season. With one package, 100 affected persons can be treated up to five days in the aftermath of a hurricane. Unused items are left for hospitals' everyday operations and are not taken back by Direct Relief. If it is predictable that a hurricane will strike a particular region, Direct Relief can provide additional packages at short notice. Such pre-positioning activities can be decisive for the success of relief operations. Therefore, the preparation phase is often in the focus of practitioners and researchers likewise [Kovács and Spens, 2007].

- Response Phase
  Shortly after a disaster, preparation plans are carried out aiming at reducing the suffering of those affected. According to Balcik and Beamon [2008], the first days after a catastrophic event are decisive for a relief agency to successfully implement a humanitarian operation. More precisely, Mitsotakis and Kassaras [2010] indicate that the first 72 hours and Ahmadi et al. [2015] state that even the first 12 hours are of vital importance. Response activities have therefore to be performed under high time pressure.

  One of the main tasks in this phase is to deliver relief items, which have ideally been pre-positioned prior to a disaster, to those affected. Often aid items are transported to a local distribution center where they can be collected by people in need [Murali et al., 2012]. Locations of such distribution centers should be chosen to cover the majority of demand in the affected area. Especially in the case of a destroyed infrastructure it is crucial to have relief facilities within walking distance. However, injured people may not be able to move and must be treated on-site. Finding victims and providing life saving assistance belong to so-called *search and rescue* activities in the response phase [Alexander, 2002, p. 155]. Only if the patient's condition permits, transportation to a health-care facility is possible. In the case of large-scale disasters, local hospitals may be overwhelmed by the huge demand and temporary medical facilities are additionally needed. In practice, tents, public buildings or mobile hospitals can serve as operating rooms for surgeries and medical treatment [Coppola, 2011, p. 312].

Evacuation is another import issue for response officials. Since the situation in the aftermath of a disaster may be chaotic, a careful and coordinated evacuation process is highly challenging. Moreover, evacuees have to leave the danger zone as fast as possible. Shelters, e.g. school gyms or airport hangars, are offered where evacuees can stay until their homes are save for return [Coppola, 2011, p. 324]. In order to arrive at the assigned shelter, evacuees use either their own cars or coaches especially provided for this purpose [Özdamar and Ertem, 2015]. In the latter case scheduled departures and bus stops are determined to guarantee a smooth process. Due to panic and unpredictable behavior in emergency situations, evacuees taking their own cars often ignore the given escape routes, leading to potential congestions and considerable delays [Caunhye et al., 2012]. Relief managers have to take such difficulties into account and should be flexible in adapting to unexpected circumstances especially in the response phase.

- Reconstruction Phase

  In contrast to immediate response the reconstruction phase, also referred to as recovery phase, is of a more strategic nature. After providing care to the affected population, important reconstruction works begin. In particular, a functioning infrastructure is essential for a community to get back to normality as quickly as possible [Hoyos et al., 2015]. Besides rebuilding destroyed facilities and roads, new houses have to be build for families who have become homeless. In order to be better prepared for the next disaster, selected buildings, main roads and bridges should be strengthened. As a result, the phases in Figure 2.3 can also be seen as a cycle where the reconstruction phase is linked to the preparation phase.

  According to Coppola [2011, pp. 377], this phase has been poorly investigated so far, resulting in a lack of understanding and unsystematic activities of those responsible. Especially for regions facing disasters on a regular basis, far-sighted decisions and solutions are required for a society to become more resistant to upcoming hazards.

The strategies described in these phases require careful planning and efficient implementation. However, numerous challenges exist in fulfilling these tasks. The population is growing rapidly and urban areas are more and more shifted to vulnerable regions, e.g. flood- or storm-prone coasts, increasing the susceptibility to natural hazards [CRED/UNISDR, 2015]. Moreover, a multitude of problems arise during the various disaster phases, putting high demands on emergency officials. Therefore, the main hurdles on the way toward a smooth relief process are described next.

2.1.2 Challenges

Besides national and international relief organizations, many other actors like military, fire departments and local governments are involved in response activities in the aftermath of a hazard. Each of theses actors has its own corporate structure and provides specific goods or services. Coordinating the efforts of a variety of different actors is indeed challenging [Balcik et al., 2010]. The main reason may be the complex and confusing situation immediately after the disaster as well as the incomplete level of information. In particular, poor or missing communication may lead to higher costs, longer lead times [Simatupang et al., 2002] and unnecessary inventories [Davis et al., 2013]. Even language barriers can hinder a desired collaboration [Moore et al., 2003], requiring a uniform information platform. For instance, the humanitarian supply chain management system SUMA supports relief efforts concerning inventory control and coordination issues [PAHO, 2001]. However, it is not always possible to implement and use such an information system. Since human and financial resources are limited, necessary staff trainings or the participation in coordination meetings are often not affordable [Moore et al., 2003].

In general, budget restrictions limit the range of possible actions of relief organizations significantly. If funding is not excessively available, helping people in need in an adequate manner is almost impossible. For instance, costs for search and rescue operations amounted to approximately US$10.8 million in the Haiti earthquake in 2010 [Coppola, 2011, p. 311]. In addition, aid organizations spend on average US$50 billions on purchasing relief items and services [Falasca and Zobel, 2011]. Such expenditures have to be covered by different funding sources. In this regard, donations represent a main source of income and are given by, e.g., governments, national and international non-profit organizations or single individuals [Coppola, 2011, p. 388]. However, only if disasters actually occur and are covered in the media the willingness to donate increases [Van Wassenhove, 2006]. Therefore, resources are rather available in the response phase, making it difficult to implement preparation strategies as desired [Tomasini and van Wassenhove, 2009, p. 50]. In addition, donors often demand that their money is used directly for the affected people, leading to massive problems concerning fundraising for proactive actions like trainings and preparation investments [Toyasaki and Wakolbinger, 2014]. As a result, relief organizations have to pay attention to their expenses and costs, restraining their ability to act especially in the preparedness phase.

In addition to financial contributions, in-kind donations are very popular but can cause a multitude of problems. For instance, medical donations may be expired, contaminated or in a foreign language which no recipient can read [Alexander, 2002, p. 199]. In addition, useless or unsolicited goods can clog distribution channels, leading to delayed relief assistance [Van Wassenhove, 2006]. Relief organizations even go so far as to refuse unwanted and needless donations in order to preserve the functionality of their supply chain.

Besides the aforementioned challenges the uncertain nature of disasters represents the main hurdle in achieving a smooth and efficient relief response. In particular, neither the exact location nor the timing or the scale of a disaster are precisely known in advance. Especially in the case of sudden-onset disasters like earthquakes the degree of devastation is hardly predictable. Even if floods and storms can be forecasted, considerable uncertainties remain: How much of what kind of relief item will be needed? Which roads will be damaged and become impassable? In addition, reliable estimations concerning financial and human resources in the aftermath of a disaster are extremely difficult [Hoyos et al., 2015]. In an emergency situation circumstances are chaotic and humans may react unexpectedly. Due to such uncertain behavior, relief plans may be impossible to realize [Caunhye et al., 2012]. Taking the high degree of uncertainty into account, decisions have to be made without reliable and full information. Therefore, emergency managers have to operate and decide in difficult conditions where the environment is changing rapidly. In such circumstances, defining suitable scenarios is a preferred method to capture the stochastic nature of disasters [Hoyos et al., 2015].

2.1.3 Scenario Definition in Disaster Management

In general, a scenario represents a possible and plausible situation or event in the future [Schnaars, 1987]. In emergency settings, a good description of scenarios is given by Alexander [2002, p. 42]: "*In the context of disasters, a scenario is a hypothetical progression of circumstances and events designed to illustrate the consequences of some decision, action or impact. It is usually designed to provide a logical answer to the question "What would happen if...?"*." Often past disasters like a historical earthquake are used to define potential consequences of future earthquakes near the same epicenter. Possible impacts on, e.g., infrastructure, response activities and aftershocks can be reflected by appropriate scenario definitions.

It is an advantage that scenarios represent a set of alternative future developments where experience and expert knowledge are incorporated. In this sense, defining scenarios can even help aid agencies to identify potential steps which can be taken before the disaster occurrence [Bañuls et al., 2012].

Rottkemper and Fischer [2013] describe a general scenario generation process in humanitarian contexts. In particular, the authors generate scenarios for the case of a malaria outbreak in Central Africa.

In the humanitarian literature discussed in the following, "*each scenario contains information on demand for relief goods and, optionally, on potential damage to inventories and/or transportation links, i.e. a scenario is a data set containing the relevant information on these aspects.*"[Grass and Fischer, 2016a]. As an example, assume that scenarios for a predicted flood should be developed. According to the weather report, the flood will probably hit a specific region. Based on past efforts to strengthen the infrastructure in this area, the emergency planner has to estimate the corresponding degree of destruction. In particular, it should be identified which roads and/or bridges may be resilient enough to withstand the disaster and can be therefore used for relief purposes in the response phase. In addition, experience related to past floods may help to define the type and quantity of relief items demanded during and after the flooding. When information becomes available during a humanitarian operation, scenarios may have to be adapted very quickly. Comes et al. [2012] propose an efficient technique for scenario updates in emergency situations.

Different scenarios are not necessarily equally likely. Therefore, a probability of occurrence is often assigned to each scenario. The more likely a scenario will occur the higher its corresponding probability. However, rare events with a low probability of occurrence, but with catastrophic consequences play a crucial role in the humanitarian context. Although not very likely, such scenarios should be taken into account while planning for disasters. Determining probability distributions may be based on historical data and/or available forecasts. If data is scarce, experts can be consulted to estimate the required probabilities. As stated by Salmerón and Apte [2010], defining scenarios and probabilities of occurrence may be a delicate task.

2.2 Quantitative Models in Disaster Management: A Literature Review

As described above, relief officials are faced with a complex and chaotic situation in the event of disaster where decisions have to be made under a high degree of uncertainty. Quantitative optimization models, which are able to incorporate uncertainties, can support the decision-making process substantially. The literature is therefore rich in this regard. Selected approaches for modeling uncertainties are described in the following.

2.2.1 Two-Stage Stochastic Programs

While different approaches for embedding uncertainties into a model exist in the literature, two-stage stochastic programming is very popular in disaster management. This is because uncertainties can be incorporated into a model by this approach, facilitating the decision-making process of aid organizations substantially. In particular, decisions are made in two stages. On the first stage, decisions have to be taken when future developments are uncertain. After the realization of a random event, second-stage decisions can be made. An example of a two-stage decision process is shown in Figure 2.6 where a hurricane represents the random event.

Based on a specific hurricane forecast, demand and damage potential can be estimated (input data in Figure 2.6). These estimations as well as purchasing and transportation costs influence the first-stage decisions concerning the procurement and storage of relief items prior to the hurricane's landfall, where the final outcome is not yet known. When the hurricane has hit a specific region, i.e. the random event has occurred, actual demand becomes known. Based on this information the pre-positioned items can now be delivered to the affected population. Moreover, pre-positioning decisions made on the first stage can be adjusted to the actual situation by redistributing relief items to facilities where needed. Decisions concerning transportation quantities and destinations belong to the second stage, also referred to as the *recourse stage*.

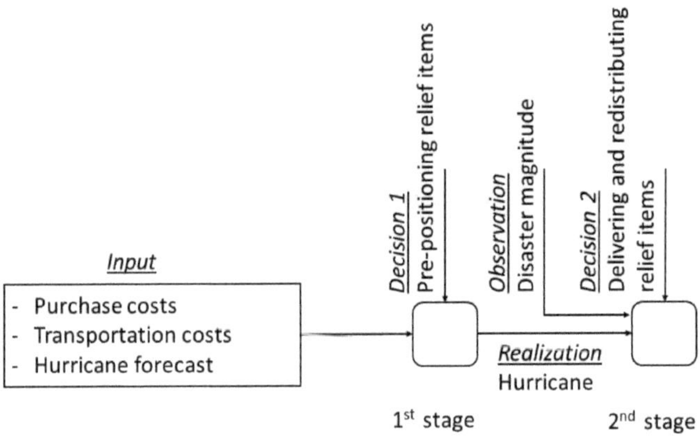

Figure 2.6: Decisions in a two-stage stochastic problem for a hurricane example (based on Schade [2012, p. 46]).

It should be mentioned that Figure 2.6 illustrates a situation where first-stage decisions are made before and second-stage decisions after the occurrence of a disaster. It is also possible to have cases where both decision types are made in the aftermath of a disaster. For instance, sudden-onset disasters like earthquakes are hardly predictable. Therefore, after the occurrence of an earthquake first-stage decisions could be to deliver relief items to the affected area where demand is initially estimated based on the earthquake's magnitude. However, actual demand and network conditions are unknown shortly after the disaster. When the situation has calmed down information on these aspects becomes available, representing the *Realization* in Figure 2.6. After observing the entire extent of the damage, ordering additional and most required items could belong to the second-stage decisions.

As described in the previous section, scenarios are often used to describe possible future outcomes. For instance, wind speed of the predicted hurricane may change in the next few days leading to different demand realizations, i.e. scenarios. Two-stage stochastic programming can take such uncertainties into account in the form of scenarios, supporting relief agencies in their difficult decision-making process. Note that, in general, only first-stage decisions are actually realized whereas second-stage decisions are rather provisional. The reason is that second-stage decisions are valid for a specific scenario and are therefore only executed after its realization. However, it is very likely that the actual disaster outcome will deviate from pre-defined scenarios to a certain extent, requiring new or adjusted transportation decisions in the aftermath. Nevertheless, provisional second-stage decisions help to decide, e.g., how many items to purchase on the first stage in order to provide sufficient relief supplies on the second stage.

An overview of two-stage stochastic programs in disaster management is given in the following.[1] In particular, Table 2.1[2] summarizes publications all proposing two-stage stochastic programs for the pre- and the post-disaster phase and for different types of disasters sorted by publication year. The list shows intense publication activities since the year 2010, i.e. the year in which Rawls and Turnquist published their seminal paper on pre-positioning.[3]

As shown in column "Application Case" in Table 2.1, hurricanes and earthquakes are the most often used numerical examples in the relevant literature. This is in contrast to the fact that floods belong to the most common disasters (see Figure 2.1). Nevertheless, relatively few publications address the specific characteristics of flood disasters, e.g. Chang et al. [2007] and Wang and Huang [2013]. However, the models summarized in Table 2.1 are generally applicable for every kind of disaster. Some authors also generate random examples without specifying a concrete disaster setting, namely Fan and Liu [2010], Bozorgi-Amiri et al. [2012], Tricoire et al. [2012] and Klibi et al. [2013].

Besides different disaster types, publications can be distinguished according to the point in time when the relevant decisions, namely the first-stage decisions, are made (see also the column "Disaster Phase" in the table above). In particular, first-stage decisions are either made before a disaster occurs (pre-disaster phase) or in the aftermath (post-disaster phase).

Pre-Disaster Phase

As shown in Table 2.1, almost all of the models considered here belong to the pre-disaster phase, since two-stage stochastic programming is particularly well suited for disaster preparation. Decisions concerning, e.g., evacuation routes and shelter locations have to be made in the preparation phase (on the first stage), not knowing the actual number of affected people, their demand and conditions of infrastructure in advance. As described in Section 2.1.1 such preparation decisions cannot be postponed until the disaster because otherwise relief assistance may arrive to late. Nevertheless, possible outcomes of disasters (on the second stage) can be anticipated and taken into account within the scope of two-stage stochastic programs. Therefore, this type of models is particularly popular in the pre-disaster phase.

Table 2.2[4] compares different objective functions and decisions classified according to the first and second stage.

[1]The following text in this subsection is partially taken from Grass and Fischer [2016b].
[2]The table is taken from Grass and Fischer [2016b].
[3]The corresponding problem description and mathematical formulation of the Rawls and Turnquist [2010] model is given in Section 2.3.
[4]The table is taken from Grass and Fischer [2016b].

Table 2.1: Publications classified according to disaster phase and type

No	Author(s)	Year	Title	Disaster Phase		Application Case		
				Pre-Disaster	Post Disaster	Earthquake	Hurricane	Flood
1	Barbarosoglu and Arda	2004	A two-stage stochastic programming planning framework for transportation planning in disaster response	x		x		
2	Chang et al.	2007	A scenario planning approach for the flood emergency logistics preparation problem under uncertainty	x				x
3	Balcik and Beamon	2008	Facility location in humanitarian relief	x		x		
4	Andreas and Smith	2009	Decomposition Algorithms for the Design of a Nonsimultaneous Capacitated Evacuation Tree Network	x				
5	Liu and Huang	2009	Dual-interval two-stage optimization for flood management and risk analyses	x				x
6	Liu et al.	2009	A two-stage stochastic programming model for transportation network protection	x		x		
7	Ng and Waller	2009	The evacuation optimal network design problem: model formulation and comparisons	x				
8	Fan and Liu	2010	Solving stochastic transportation network protection problems using the progressive hedging-based method	x				
9	Mete and Zabinsky	2010	Stochastic optimization of medical supply location and distribution in disaster management	x		x		
10	Peeta et al.	2010	Pre-disaster investment decisions for strengthening a highway network	x		x		
11	Rawls and Turnquist	2010	Pre-positioning planning for emergency response with service quality constraints	x			x	
12	Salmeron and Apte	2010	Stochastic optimization for natural disaster asset	x			x	
13	Cai et al.	2011	Identification of optimal strategies for improving eco-resilience to floods in ecologically vulnerable regions of a wetland	x				x
14	Falasca and Zobel	2011	A two-stage procurement model for humanitarian relief supply chains		x	x		
15	Li et al.	2011	Sheltering network planning and management with a case in the Gulf Coast region	x			x	
16	Rawls and Turnquist	2011	Pre-positioning planning for disaster response with	x			x	

(continued on the next page)

Table 2.1: Publications classified according to disaster phase and type (continued)

No	Author(s)	Year	Title	Disaster Phase Pre-Disaster	Post Disaster	Application Case Earthquake	Hurricane	Flood
			service quality constraints					
17	Bozorgi-Amiri et al.	2012	A modified particle swarm optimization for disaster relief logistics under uncertain environment	x				
18	Döyen et al.	2012	A two-echelon stochastic facility location model for humanitarian relief logistics	x		x		
19	Li et al.	2012	Shelter location and transportation planning under hurricane conditions	x			x	
20	Lodree et al.	2012	Pre-positioning hurricane supplies in a commercial supply	x			x	
21	Noyan	2012	Risk-averse two-stage stochastic programming with an application to disaster management	x			x	
22	Rawls and Turnquist	2012	Pre-positioning and dynamic delivery planning for short-term response following a natural disaster	x			x	
23	Tricoire et al.	2012	The bi-objective stochastic covering tour problem	x				
24	Wang et al.	2012	A two-stage stochastic programming model for emergency resource storage region diversion	x		x		
25	Bozorgi-Amiri et al.	2013	A multi-objective robust stochastic programming model for disaster relief logistics under uncertainty	x		x		
26	Davis et al.	2013	Inventory planning and coordination in disaster relief efforts	x			x	
27	Klibi et al.	2013	Prepositioning emergency supplies to support disaster relief	x				
28	Maqsood and Huang	2013	A dual two-stage stochastic model for flood management with inexact-integer analysis under multiple uncertainties	x				x
29	Wang and Huang	2013	A two-stage mixed-integer fuzzy programming with interval-valued membership functions approach for flood diversion	x				x
30	Ahmadi et al.	2015	A humanitarian logistics model for disaster relief operation considering network failure and standard relief time: A case study on San Francisco	x		x		
31	Hong et al.	2015	Stochastic network design for disaster preparedness	x			x	
32	Hu et al.	2015	A scenario planning approach for pre-positioning rescue centers for urban waterlog disasters	x				x
33	Verma and Gaukler	2015	Pre-positioning disaster response facilities at safe	x		x		

(continued on the next page)

Table 2.1: Publications classified according to disaster phase and type (continued)

No	Author(s)	Year	Title	Disaster Phase Pre-Disaster	 Post Disaster	Application Case Earthquake	 Hurricane	 Flood
			locations: An evaluation of deterministic and stochastic modeling					
34	Zolfaghari and Peyghaleh	2015	Implementation of equity in resource allocation for regional earthquake mitigation using two-stage stochastic programming	x		x		
35	Alem et al.	2016	Stochastic network models for logistics planning in disaster relief	x				x
36	Bai	2016	Two-stage multi-objective optimization for emergency supplies allocation problem under integrated uncertainty	x		x		
37	Noyan et al.	2016	A stochastic optimization model for designing last mile relief networks		x	x		
38	Rath et al.	2016	Bi-objective stochastic programming models for determining depot locations in disaster relief operations		x	x		
39	Rezaei-Malek et al.	2016	An approximation approach to a trade-off among efficiency, efficacy, and balance for relief pre-positioning in disaster management	x		x		
40	Tofighi et al.	2016	Humanitarian logistics network design under mixed uncertainty	x		x		

Table 2.2: Objective functions and decisions for first and second stage in the pre-disaster phase

Author(s)	Year	Objectives		Decisions	
		First-Stage	Second-Stage	First-Stage	Second-Stage
Barbarosoglu and Arda	2004	costs (transportation; switching modes)	costs (transport; shortage; inventory; switching modes)	transport	transport; unused inventory; shortage
Chang et al.	2007	costs (location; procurement; maintenance)	costs (transport; shortage)	locations; pre-positioned items	transport; unused inventory; shortage
Balcik and Beamon	2008	–	satisfied demand	locations; pre-positioned items	demand satisfaction
Andreas and Smith	2009	–	time-dependent penalty	evacuation tree	evacuation flows
Liu and Huang	2009	costs (flood diversion)	costs (excess flood)	flood diversion	amount of excess flood
Liu et al.	2009	–	costs (repair; weighted flow)	retrofitting bridges	link flow
Ng and Waller	2009	costs (capacity expansion)	evacuation time	capacity expansion	evacuation flows
Fan and Liu	2010	–	costs (repair; travel delay; unsatisfied travel demand)	link protection	link flow
Mete and Zabinsky	2010	costs (location)	costs (shortage); weighted transportation time	locations; pre-positioned items	transport; shortage
Peeta et al.	2010	expected number of broken links	costs (transport)	retrofitting links	transport
Rawls and Turnquist	2010	costs (location; procurement)	costs (transport; shortage; excess)	locations; pre-positioned items	transport; unused inventory; shortage
Salmeron and Apte	2010	–	casualties; unmet transfer population	capacity expansion (for critical/ transfer population, items and ramp space)	vehicles; transport (persons + items); shortage; relief workers
Cai et al.	2011	costs (capacity)	costs (excess flood)	flood-resisting level	amount of excess flood
Li et al.	2011	costs (location; capacity; inventory)	costs (transport; excess; shortage)	locations; capacity; resources	number of evacuees; transport; unused inventory; shortage
Rawls and Turnquist	2011	costs (location; procurement)	costs (transport; shortage; excess)	locations; pre-positioned items	transport; unused inventory; shortage
Bozorgi-Amiri et al.	2012	costs (location; procurement; transport)	costs (procurement; transport; inventory; shortage); variability of total costs	locations; pre-positioned items	transport; inventory; shortage

(continued on the next page)

Table 2.2: Objective functions and decisions for first and second stage in the pre-disaster phase (continued)

Author(s)	Year	Objectives		Decisions	
		First-Stage	Second-Stage	First-Stage	Second-Stage
Döyen et al.	2012	costs (location; holding)	costs (location; shortage; transport)	locations; pre-positioned items	transport; location; shortage
Li et al.	2012	–	unmet shelter demand; travel time	locations	shelters; number of evacuees
Lodree et al.	2012	costs (preparing for shipment; transport)	costs (transport; production)	pre-positioned items	transport
Noyan	2012	costs (location; procurement)	costs (transport; shortage; excess)	locations; pre-positioned items	transport; shortage
Rawls and Turnquist	2012	costs (location; procurement)	costs (transport; shortage; excess)	locations; pre-positioned items	transport; unused inventory; shortage
Tricoire et al.	2012	costs (location; transport)	uncovered demand	allocation of villages and vehicles to DC	transport
Wang et al.	2012	–	costs (shortage; transport)	allocation to region	shortage; transport
Bozorgi-Amiri et al.	2013	costs (location; procurement; transport)	costs (procurement; transport; inventory; shortage); variability of total costs; robustness; maximum shortage	locations; pre-positioned items	transport; inventory; shortage
Davis et al.	2013	costs (distance)	costs (transport; shortage; damaged items)	pre-positioned items; transport; unutilized capacity	transport; shortage
Klibi et al.	2013	–	costs (procurement; transport)	locations; pre-positioned items	transport; unused inventory; shortage
Maqsood and Huang	2013	costs (flood diversion)	costs (excess flood; capacity)	flood diversion	amount of excess flood; capacity expansion
Wang and Huang	2013	costs (flood diversion)	costs (excess flood; capacity)	flood diversion	amount of excess flood; capacity expansion
Ahmadi et al.	2015	costs (shortage; location); distribution time	costs (violating SRT constraints)	locations; shortage; travel time	delivery time exceeding SRT
Hong et al.	2015	costs (location; pre-positioning)	costs (shortage)	locations; pre-positioned items	shortage
Hu et al.	2015	costs (location; pre-positioning)	costs (transport; delay; inventory; shortage; excess)	locations; pre-positioned pumps	transport; delayed time; shortage; unused inventory

(continued on the next page)

Table 2.2: Objective functions and decisions for first and second stage in the pre-disaster phase (continued)

Author(s)	Year	Objectives		Decisions	
		First-Stage	Second-Stage	First-Stage	Second-Stage
Verma and Gaukler	2015	–	costs (transport)	locations	transport
Zolfaghari and Peyghaleh	2015	costs (retrofitting)	costs (reconstruction)	retrofitting buildings	reconstruction expenditures
Alem et al.	2016	costs (pre-positioning; vehicles)	costs (inventory; shortage)	pre-positioning items; vehicles	inventory; shortage; additional procurement; vehicles; transport; budget
Bai	2016	–	unmet demand; travel time; costs (vehicle; transport)	suppliers; vehicles	transport
Rezaei-Malek et al.	2016	costs (location; pre-positioning)	costs (transport; excess); utility level; imbalance measure	locations; pre-positioned items	transport; unused inventory; shortage; utility level; levels among demand points
Tofighi et al.	2016	costs (location; inventory)	distribution time; weighted travel time; costs (shortage; excess)	locations; pre-positioned items	transport; unused inventory; maximum transportation time

Objectives

The proposed models can be distinguished primarily based on their objective functions. As stated in Table 2.2, minimization of first- and second-stage costs is most commonly used since relief organizations are generally reliant on donations or other financial grants. As mentioned in Section 2.1.2 donations are often received in the aftermath of a disaster and earmarked for relief purposes only. This makes it considerably difficult to finance preparation strategies and long-term projects. Hence, costs play a crucial role for relief organizations with restricted budgets especially in the preparation phase. Nevertheless, meeting demand in a humanitarian setting should be prioritized due to far-reaching consequences with respect to human lives. Often, penalty costs for unmet demand (shortages) are included in the second-stage objective to minimize unsatisfied demand, e.g. Barbarosoğlu and Arda [2004], Wang et al. [2012], Bozorgi-Amiri et al. [2012], Rawls and Turnquist [2010]. A few authors address the fulfillment of demand explicitly by maximizing satisfied demand [Balcik and Beamon, 2008], by minimizing casualties [Salmerón and Apte, 2010] or maximizing accessibility [Noyan et al., 2016]. Others, e.g. Mete and Zabinsky [2010], Tofighi et al. [2016], incorporate multiple objectives measured in different units, like minimization of costs (monetary units) and delivery time (time units), into their model.

In some cases, only second-stage objectives are formulated. For instance, first-stage decisions in Klibi et al. [2013] are made on locations and pre-positioning while taking budget restrictions into account instead of formulating a cost minimization objective on this stage.

Conflicting objectives, like minimization of costs on the one hand and maximization of satisfied demand on the other, are common in disaster management and deserve special attention. As shown in Table 2.2 almost all of the publications make a compromise by minimizing operating costs and minimizing unmet demand, introducing penalty costs. However, it can be very difficult to quantify such penalty costs in practice. Hong et al. [2015] avoid the use of penalty costs for shortages by introducing chance constraints where demand has to be fulfilled with a predefined probability. Constraints which have to be satisfied with a specific user-defined probability are called chance or probabilistic constraints [Birge and Louveaux, 2011, Ch.3.2]. Hong et al. [2015] show that their model leads to a higher reliability level than the penalty cost approach proposed by Rawls and Turnquist [2010]. Therefore, researchers should continue to develop alternative approaches for solving multi-objective models and develop new ways to handle critical input parameters which cannot easily be estimated. A recent survey on multi-objective approaches in humanitarian aid can be found in Gutjahr and Nolz [2016].

Decisions

Besides the objective functions, the decisions to be made are a characteristic feature of the models. They are separated in Table 2.2 into first- and second-stage decisions. As mentioned above, this differentiation is important since, in general, only the first-stage decisions are actually executed and therefore they represent the most relevant part of a two-stage stochastic program. Even though a model may provide, for example, transportation quantities of commodities for each scenario, it is highly unlikely that exactly one of the scenarios that was taken into account beforehand will actually be realized. Hence, also the first-stage decisions do not have to be optimal ex post, but still may be the best decision that could be made ex ante. As indicated by the number of models with first-stage decisions "pre-positioned items" in Table 2.2, in-advance storage of relief items is the most important preparation strategy. A few authors focus on other first-stage decisions like retrofitting of links [Fan and Liu, 2010, Peeta et al., 2010], buildings [Zolfaghari and Peyghaleh, 2015] or bridges [Liu et al., 2009], capacity expansions [Ng and Waller, 2009, Salmerón and Apte, 2010] and travel time [Ahmadi et al., 2015]. In the context of flood disasters, Wang and Huang [2013] focus solely on the design of diversion plans. When a river reaches a critical level, floodwater is redirected to flood retention regions, mitigating flooding of urban areas.

Table 2.2 additionally shows the second-stage decisions, where the transport of commodities in the aftermath of a disaster is the most popular aspect to be modeled. This is reasonable since delivering aid supplies is a crucial task in the response phase (see Section 2.1.1). For instance, Mete and Zabinsky [2010] propose a two-stage program, where commodities are pre-positioned in warehouses on the first stage and the quantities to be transported are selected on the second stage. In addition, a separate transportation plan for last-mile distribution after an earthquake is given which uses the results from the corresponding two-stage program.

Since evacuation is an essential task in disaster management, several authors formulate evacuation plans on the second stage, i.e. after disaster occurrence, e.g. Salmerón and Apte [2010] and Li et al. [2012, 2011]. Andreas and Smith [2009] decide on the first stage if a specific link is included in the evacuation routes, designing an evacuation tree network. Such an evacuation tree can be used, for example, to place arrow symbols in buildings to guide evacuees.

However, the structure of two-stage stochastic programming would allow the decision maker to define an evacuation plan on the first stage with refinements on the second stage. For instance, the population which will most likely be hit by a hurricane can be evacuated to shelters with appropriate capacity. Since paths of hurricanes can change significantly with changing weather conditions, possible additional second-stage evacuations can be taken into account on the first stage. However, no such attempt has been made in the two-stage stochastic literature so far.

In Table 2.3[5], uncertainties, problematic assumptions and special features of the different models for the pre-disaster phase are summarized.

Table 2.3: Uncertainties, problematic assumptions and special features of pre-disaster models

Author(s)	Year	Uncertainty on the First-Stage	Problematic Assumptions	Special Features
Barbarosoglu and Arda	2004	demand; arc capacity; supply		inter-modal

(continued on the next page)

[5]The table is taken from Grass and Fischer [2016b].

Table 2.3: Uncertainties, problematic assumptions and special features of pre-disaster models (continued)

Author(s)	Year	Uncertainty on the First-Stage	Problematic Assumptions	Special Features
Chang et al.	2007	demand	no damage to links and facilities	multi-level of rescue facilities (local rescue base; regional and head center)
Balcik and Beamon	2008	demand; transport costs; time to satisfy demand	no damage to facilities; single demand point	
Andreas and Smith	2009	number of evacuees; link capacity; travel time	single destination point	
Liu and Huang	2009	flood flow; costs		interval-parameter programming
Liu et al.	2009	transportation links	no damage of retrofitted links; link flow is known	bridge retrofitting
Ng and Waller	2009	number of evacuees; shelter and link capacity; travel time		
Fan and Liu	2010	links; travel demand; travel time	no damage of protected links	link protection
Mete and Zabinsky	2010	demand; transportation time; penalty costs	no damage to facilities	transportation plan produced by two-stage program used for updating routing plan
Peeta et al.	2010	transportation links	failure probabilities of links are known; independent link failures	network reliability
Rawls and Turnquist	2010	demand; inventory; transportation links		
Salmeron and Apte	2010	demand; health conditions of affected population; population survival rate; transportation time; relief workers	no damage to facilities	priority affected
Cai et al.	2011	flood flow		fuzzy interval-stochastic programming
Li et al.	2011	evacuees; costs (transport; evacuees)	no damage to shelters	
Rawls and Turnquist	2011	demand; inventory; transportation links		reliability constraints
Bozorgi-Amiri et al.	2012	demand; costs (procurement; transport); inventory		
Döyen et al.	2012	demand; capacity; transportation time; costs (transport; location)	uncapacitated and undamaged first-stage facilities	two-echelon network
Li et al.	2012	demand; shelter; accessibility; travel time	private vehicles available	evacuees' route choice behavior
Lodree et al.	2012	demand	no damage to facilities and links	
Noyan	2012	demand; inventory; transportation links		cVaR

(continued on the next page)

Table 2.3: Uncertainties, problematic assumptions and special features of pre-disaster models (continued)

Author(s)	Year	Uncertainty on the First-Stage	Problematic Assumptions	Special Features
Rawls and Turnquist	2012	demand (type and location); supply reliability		dynamic allocation (short-term demand)
Tricoire et al.	2012	demand	no damage to facilities and links	
Wang et al.	2012	demand	no damage to links	
Bozorgi-Amiri et al.	2013	demand; costs (procurement; transport); inventory		robust two-stage stochastic program
Davis et al.	2013	demand; travel speed (miles per hour) based on congestion		traffic congestion; equity in service; inventory coordination
Klibi et al.	2013	demand; facility and vendor's inventory damage; transportation links	sufficient capacity of DC	multi disasters; three-echelon system
Maqsood and Huang	2013	flood flow; retention capacities; costs; diversion targets		interval-parameter programming
Wang and Huang	2013	flood flow		fuzzy programming with interval-valued membership functions
Ahmadi et al.	2015	travel time	no damage to facilities; demand is known	standard relief time (SRT)
Hong et al.	2015	demand; arc capacity	no damage to facilities	chance-constraints
Hu et al.	2015	demand	no damage to facilities and links	multi-attribute utility function
Verma and Gaukler	2015	demand; inventory (uniform and normal distribution)	no damage to transportation links	distance-dependent destruction level
Zolfaghari and Peyghaleh	2015	damage of buildings		equity constraints
Alem et al.	2016	demand; incoming supplies; budget; inventory; links		dynamic multi-period horizon; minimax-regret, semi-deviation and cVaR
Bai	2016	demand; supplier's inventory; links	normally distributed supply	fuzzy random variables and chance constraints
Rezaei-Malek et al.	2016	demand; demand priority; inventory; transportation time; tolerable shortage		imbalance and efficacy measures
Tofighi et al.	2016	demand; costs; travel times; inventory		critical and non-critical items; priority of demand; two-echelon relief chain; possibilistic programming

Uncertainties and Problematic Assumptions

Since disasters are often unpredictable, different kinds of uncertainties have to be taken into account on the first stage (see Table 2.3). Besides demand and the condition of facilities and/or roads, also costs, e.g. for transportation, may be uncertain; see Bozorgi-Amiri et al. [2012]; Li et al. [2011]; Döyen et al. [2012]; Balcik and Beamon [2008]. However, in some models it is assumed that commodities stored in advance are readily available after a disaster, i.e. that facilities as well as the links remain undamaged, e.g. Chang et al. [2007]; Lodree et al. [2012]. As it is highly unlikely that, for example, roads and/or warehouses are in the same condition as before, these assumptions are rather unrealistic and can be viewed as problematic (see column "Problematic Assumptions" in Table 2.3). Considering network reliability assuming the independence of link destruction, as in Peeta et al. [2010], can also be regarded as critical especially in the humanitarian context. Moreover, Peeta et al. [2010] require knowledge of the damage probability of a link in advance, but such determinations are challenging in practice. Another unrealistic assumption is made by Liu et al. [2009] where retrofitted bridges cannot be destroyed by an earthquake.

In the model proposed by Balcik and Beamon [2008] demand arises at only one location. However, since catastrophic events can devastate entire regions this will usually not be the case.

Such unrealistic assumptions are often made to simplify the model's complexity, leading at the same time to a limited utility of the proposed model. Due to the widespread destruction resulting from disasters, assuming undamaged links and facilities is indeed problematic. Hence, the more realistic scenarios are defined, the more valuable are the results for practitioners.

Special Features

In Table 2.3, also special features are presented to distinguish the models from each other and to highlight the innovative character of a specific model. Bozorgi-Amiri et al. [2013] formulate an objective with a robustness measure where infeasible solutions are penalized by a penalty parameter. Alem et al. [2016] compare different risk measures, namely the minimax-regret model, the semi-deviation measure and the conditional Value-at-Risk (cVaR). The cVaR concept at the confidence level β has its origin in financial applications and represents, in the context considered here, the mean costs under the condition that one of the $(100 - \beta)\%$ worst-case events occurs. The authors show that the cVaR concept leads to higher demand satisfaction than the minimax-regret model and the semi-deviation measure. The latter is calculated in the same way as the statistical standard deviation, but only values below the average are taken into account. The minimax-regret approach tries to reduce the worst-case regret as much as possible. Alem et al. [2016] define regret as the second-stage costs minus the optimal costs for the corresponding scenario when a wait-and-see problem is considered.

As already mentioned, the first 72 hours after a disaster are often defined as the critical time period for relief operations [Mitsotakis and Kassaras, 2010]. Rawls and Turnquist [2012] address this issue by extending their former work such that short-term demand in this time frame has to be satisfied at specific emergency shelters. Other authors argue that especially the first 12 hours, the so-called *standard relief time*, are most critical [Ahmadi et al., 2015]. The authors use penalty costs to avoid violations of these time constraints, ensuring a faster delivery. Their model can also be used in the post-disaster phase by temporarily setting up facilities when demand and inventory levels are already known but the network conditions are still uncertain. In this regard, the approach of Ahmadi et al. [2015] is more flexible, increasing the added value of their model.

Besides fast delivery of first-aid supplies, a fair distribution of these supplies and medical care represents a key principle of humanitarian operations. For instance, Davis et al. [2013] take negative congestion effects into account when developing an equitable distribution of first-aid supplies. The equity concept is also included in the model proposed by Zolfaghari and Peyghaleh [2015] through the use of specific constraints. Salmerón and Apte [2010] prioritize affected populations, with critically wounded people evacuated first. The authors attempt to ensure a fair treatment of the affected persons by taking different severity of injuries into account. Hu et al. [2015] apply multi-attribute utility theory to measure the utility of, for example, rescue centers. Here, used resources are set in relation to available resources and to demand of the affected population in order to measure demand satisfaction. Based on the time, priority and demand-weighted utility level of the transported relief items, Rezaei-Malek et al. [2016] introduce additional imbalance and efficacy measures. For instance, the imbalance of the network is measured by the maximum utility difference between various demand points. Accordingly, the model proposed by Rezaei-Malek et al. [2016] leads to a more balanced demand satisfaction and to a fairer distribution.

Li et al. [2012] propose a bi-level program, where a two-stage stochastic program is formulated on the upper level and the route choices of evacuees are considered on the lower level. In particular, decisions on shelter locations and the number of evacuees allocated to these shelters belong to the upper level and are made by facility planners, e.g. relief organizations. The lower level contains the routing decisions of the network users, i.e. the evacuees, who try to leave the affected area as quickly as possible.

Other features, as stated in Table 2.3, refer to multi-echelon supply chains, e.g. Döyen et al. [2012] and Klibi et al. [2013], since relief chains comprise, in general, several members. Such an extension is necessary for establishing a more realistic counterpart to the single-echelon model as one- or two-echelon networks represent reality only in a limited way.

Klibi et al. [2013] model the possibility of several disasters in consecutive periods within the two-stage structure. This is motivated by real events like the Fukushima nuclear disaster in 2011, where an earthquake caused a tsunami, leading in turn to damage to power stations and finally to three nuclear meltdowns. Even relatively light aftershocks after an earthquake may lead to additional destruction, highlighting the importance of the approach of Klibi et al. [2013]. However, two-stage programming can also be expanded to multiple stages, where additional information concerning, e.g., possible aftershocks or outbreaks of epidemics after disastrous events can be incorporated in the model formulation, allowing a more detailed decision process. For instance, Rennemo et al. [2014] propose a three-stage stochastic program, where information regarding the actual amount of supplies and the number of vehicles is available on the second stage, and the destruction level of the infrastructure becomes known on the third stage only. Dupačová et al. [2000] give a survey of different scenario tree generation techniques, which could be useful for such multi-stage stochastic programs.

In general, scenarios are used to model uncertainties, and the respective probabilities of occurrence are often derived from historical data. However, parameters *within* a scenario can also be highly uncertain, as stated in Tofighi et al. [2016]. This is due to the uniqueness of each disaster, which renders the determination of probability distributions problematic. For instance, although it is possible to predict how likely it is that a hurricane will hit a specific part of the American southeast coast, the actual demand for relief goods for this hurricane scenario may remain uncertain. Hence, Tofighi et al. [2016] combine the scenario-based approach with possibilistic distributions, where imprecise data is modeled through the use of fuzzy numbers.[6] Along a similar line, Cai et al. [2011] as well as Wang and Huang [2013] apply the fuzzy programming technique to their flood diversion models. Bai [2016] integrates fuzzy random variables into a chance constraint formulation.

Another approach to cope with uncertainties if appropriate probability distributions are difficult to define is the use of intervals as proposed by Liu and Huang [2009] and Maqsood and Huang [2013]. They assume that values of uncertain parameters lie within specific interval limits. In this way, the corresponding solutions are also within an interval, allowing the decision maker to choose from a variety of decision alternatives instead of having to accept one specific "optimal" solution. In order to avoid the difficult determination of probability distributions, Caunhye et al. [2016] minimize the total response time for each scenario in their pre-positioning location-routing model; then the solution for the worst case is chosen. In this way, the proposed model provides a preparation plan for the worst case neglecting the probability of occurrence for all other scenarios. Since neither the expected value nor probability distributions are used by Caunhye et al. [2016], the respective model does not fit into the classical two-stage stochastic framework and is therefore not included in the overview tables above.

As mentioned in Section 2.1.2, relief organizations do not act by themselves in emergency situations but rather as one unit among others like military units, governmental and non-governmental actors. Their objectives may be in conflict with each other such that decisions taken by one party have to take other stakeholders into account. Uncoordinated decisions may lead to unnecessary duplications of inventories and/or distributions which may interfere with each other. Currently, almost no work within the two-stage stochastic programming context has been devoted to integrating different actors in humanitarian supply chains into the model-building and decision-making process. One exception is the publication of Davis et al. [2013] where resources of different participants are coordinated.

Post-Disaster Phase

In the publications considered in this section, first- and second-stage decisions are both made *after* the disaster. Although the planning in two stages is reasonable when the disaster has not yet occurred, two-stage stochastic programming may also be valuable

[6] For a detailed description and a thorough introduction to the fuzzy concept see for example Zhang and Liu [2006].

in the aftermath of a disaster. In the event of a catastrophe, reliable information is often difficult to obtain due to destruction to roads and the breakdown of communication systems. In such situations, shortly after disaster occurrence, decisions still have to be made under uncertainties which are resolved only later, emphasizing the need for two-stage programming in the post-disaster phase. However, only three models have been proposed for this phase (see Table 2.4 and 2.5[7]). Since the first 12 hours after a disaster are especially critical, relief actions have to be taken instantly once information starts becoming available. When sudden events like terrorist attacks occur, the determination of appropriate scenarios is nearly impossible as there is hardly enough time and knowledge available. In contrast to this, for disaster-prone regions a set of possible scenarios can be prepared in advance and hence two-stage models are better suited for these situations.

Falasca and Zobel [2011] propose a procurement model where relief items are ordered from suppliers in a concrete emergency situation. Here, purchasing decisions (first stage) are indeed made after disaster occurrence but when the environment is still in a chaotic state and actual demand is not yet known. Moreover, donations are not instantly available but rather collected during an ongoing relief operation, and hence are uncertain. When demand can be identified more accurately, the relief organization can purchase additional items (second-stage decisions), taking actual donations into account. Note that conditions of facilities and roads are not relevant for the procurement decisions in Falasca and Zobel [2011] and hence are not addressed in their model. In contrast, Rath et al. [2016] consider, among others, vehicle procurement on the first stage when road damage and hence travel times are still unknown. On the second stage, when conditions of roads etc. have become known, transport decisions are made such that demand coverage is maximized.

Table 2.4: Objective functions and decisions for first and second stage of post-disaster phase

Author(s)	Year	Objectives		Decisions	
		First-Stage	Second-Stage	First-Stage	Second-Stage
Falasca and Zobel	2011	costs (procurement)	costs (procurement; shortage)	procurement; choice of supplier	additional procurement; shortage
Noyan et al.	2016	expected accessibility	accessibility	locations; capacity	transport; demand coverage
Rath et al.	2016	costs (location; procurement; operating budget)	demand coverage	budget; locations; vehicle procurement	transport; vehicle assignment

Table 2.5: Uncertainties, problematic assumptions and special features of post-disaster models

Author(s)	Year	Uncertainty in the First-Stage	Problematic Assumption	Features
Falasca and Zobel	2011	demand; donations		
Noyan et al.	2016	demand; accessibility		accessibility; equity
Rath et al.	2016	travel time; costs (transport)	known demand and inventory	Several model variants

[7]Both tables are taken from Grass and Fischer [2016b].

The model proposed by Noyan et al. [2016] aims to locate points of distribution in the so-called last mile network[8] in such a way that every affected person has access to first-aid supplies and at the same time relief items are distributed equally among these persons. Note that the points of distribution are locations where everyone in need can collect relief packages but is responsible himself for reaching these points. Hence, decisions concerning these locations are essential to guarantee a fair distribution of first-aid supplies.

2.2.2 Pre-Positioning of Relief Items

As indicated by Table 2.2, most first-stage decisions are concerned with pre-positioning of relief items in preparation for disasters.[9] Based on the descriptions in Section 2.1.1, pre-positioning activities are essential in preparation for disasters. Especially disaster-prone areas benefit from storing items in advance as it reduces the time needed to react to an emergency situation. The example of Direct Relief [2017] highlights the fact that hospitals offer their storage space for medicine for the entire hurricane season to ensure better preparation in case of an emergency.

Moreover, relief organizations which purchase in advance can benefit from price negotiations with suppliers and from discounts due to large quantity orders [Rawls and Turnquist, 2012]. In contrast, if no emergency supplies are stored in advance, large-scale disasters lead to a sudden increase in demand for relief items and aid organizations have to purchase these items from local or global vendors whose inventory levels are usually not sufficient. This fact is confirmed by the relief agency *CARE International*, which states that two weeks are usually required to replenish its warehouses [Duran et al., 2011]. Obviously, this time frame is not acceptable in emergency situations when survival depends on the availability of aid supplies. In general, ad-hoc purchase decisions result in higher costs, longer delivery times and therefore in delayed disaster response and unsatisfied demand [Balcik and Beamon, 2008]. Due to the relevance of the first 72 hours, pre-positioning decisions are of vital importance with regard to response time and at the same time especially challenging due to uncertainties about where, when and to what extent a disaster may occur. As highlighted in the previous section, two-stage stochastic programming is a preferred technique in the pre-positioning literature. However, other methods for modeling uncertainties are also proposed in the literature and reviewed next.

Scenario-Free Pre-Positioning Models

Some authors argue that approaches based on scenarios limit the set of future realizations. If historical data are not readily available to formulate a sufficient number of scenarios that adequately represent possible outcomes, scenario-free approaches have to be found as stated in column "Approach" of Table 2.6[10]. For instance, Galindo and Batta [2013] use expected

[8]The last mile network considered in Noyan et al. [2016] contains warehouses, called local distribution centers, which send relief items to points of distribution from where the affected population is served.
[9]This subsection is partially taken from Grass and Fischer [2016a].
[10]The table is based on Grass and Fischer [2016a].

values based on forecasts to capture uncertainties in their model. In addition, Table 2.6 summarizes objectives, uncertainties and assumptions of the scenario-free approaches.

Table 2.6: Scenario-free approaches: objectives, uncertainties, approaches capturing uncertainties and assumptions

Author(s)	Year	Objective	Uncertainties	Approach	Assumptions
Campbell and Jones	2011	min costs, e.g. for restocking, salvage	demand (normal distribution)	newsvendor	no damage to facilities and transportation links; single demand point
Murali et al.	2012	max demand covered	demand (log-normal distribution)	chance-constrained	no damage to facilities and transportation links; optimal assignment of demand to facilities
Galindo and Batta	2013	min costs	demand; inventory (uniform integer distribution)	expected value	no damage to transportation links; short-term planning (2 days before disaster occurence)
Renkli and Duran	2015	min weighted distance	demand; transportation links	chance-constrained	no damage to facilities
Verma and Gaukler	2015	min transportation costs	demand; inventory (uniform and normal	distance-damage function distribution)	no damage to transportation links

The first who refrained from using scenarios in the context of aid pre-positioning were Campbell and Jones [2011], who assume normally distributed demand. Their objective is to decide where to open facilities and how many of certain relief items to store so that total costs, including for example restocking and disposal costs, are minimized. In their model demand occurs at only one specified location, given that supply facilities and hence inventories can be destroyed by a disaster with known and independent probabilities. In the situation studied by Galindo and Batta [2013], emergency items are stored two days before the landfall of a hurricane using expected values of demand based on available forecasts. However, the path and strength of a hurricane can change at short notice with changing weather conditions, such that more advanced approaches for modeling uncertainty are required. Recently, Verma and Gaukler [2015] introduced a random variable which is distance-dependent and represents the fraction of total facility capacity which is no longer available in the aftermath of a disaster. Here, the destruction level of facilities decreases with increasing distance to the disaster location, e.g. the epicenter of an earthquake. However, it is required that the underlying distribution of the random variable is known. For instance, the authors assume uniform and normal distributions of the distance-damage function. It is highly questionable whether such distributions can sufficiently reflect reality. In addition, transportation disruptions and facility damage are not considered in their model.

Uncertainty can also be captured by the chance-constrained method as in Murali et al. [2012]. Their optimization model is a variant of the maximal covering location problem, in which the amount of satisfied demand depends on the distance to a service facility. Here, constraints dealing with the assignment of relief items to demand points are required to hold with a predefined probability. Due to the fact that demand is non-negative,

Murali et al. [2012], unlike Campbell and Jones [2011], assume log-normally distributed demand. A second paper using probability constraints for pre-positioning problems has been published by Renkli and Duran [2015]. Deliveries can only take place if the distance is not too large and if the corresponding transportation link is not damaged.

The main disadvantage of continuous probability distribution approaches is that their applicability is limited; see for example the model of Campbell and Jones [2011], where only one demand point is served. In order to cope with multi-facility and multi-demand networks a common methodology is to define a finite number of scenarios within a two-stage stochastic framework as was shown in the previous section and as in the example below.

2.3 The Rawls and Turnquist [2010] Model

Although Chang et al. [2007] and Balcik and Beamon [2008] were among the first to address pre-positioning issues, researchers realized their importance only after the publication of Rawls and Turnquist [2010]. Their model is a classical two-stage stochastic program and one of the most influential pre-positioning models in the humanitarian literature. Therefore, it is used as the reference model in the following.[11]

2.3.1 Problem Description and Mathematical Formulation

For disaster-prone areas either new relief facilities have to be built or existing facilities can be used if short-term forecasts are available. In both cases, locations have to be chosen from a pool of potential locations before the occurrence of a disaster. These locations are represented by nodes in a network, as shown in Figure 2.7, where storage facilities with specific capacities can be opened and demand for aid supplies can arise.

In the aftermath of a disaster, pre-positioned items can be transported on roads, illustrated by arcs in the network in Figure 2.7. Here, the hurricane-prone Gulf- and Atlantic coast is chosen as an example where 30 nodes are used to partition the vulnerable area. As this coastal region is hit by storms and hurricanes every year it is also used for designing a large-scale case study in Section 4.1. On average 6.3 hurricanes hit parts of the Atlantic basin each year [AOML, 2016] highlighting the region's vulnerability in this regard.

The model proposed by Rawls and Turnquist [2010] determines the number of facilities and their corresponding capacities. In addition, variables for the type and quantity of relief items pre-positioned at these facilities are defined (first-stage decisions). In other words, these decisions involve locating relief facilities and storing emergency supplies before uncertainties are realized. After a disaster has occurred, the conditions of transportation links and storage facilities as well as availability of relief items are known and second-stage decisions are made. For instance, transportation decisions, i.e. how many of which type of relief items have to be transported to which demand point, belong to the second stage. If the actual disaster outcome deviates from the pre-defined scenarios, as it is often the case

[11] The following two subsections are partially taken from Grass and Fischer [2016a,b].

in practice, transportation plans can be changed in the aftermath of a disaster according to the actual needs. As mentioned above, second-stage decisions are rather provisional, but help to decide where and how much to store before the disaster occurrence in order to achieve the best possible supply afterwards.

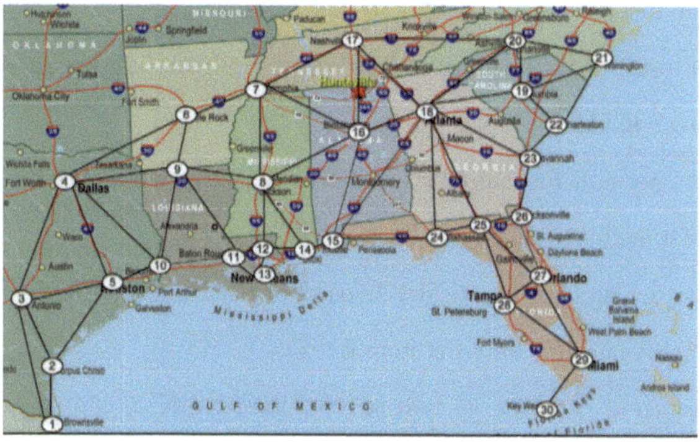

Figure 2.7: Network containing nodes and arcs using the example of the Gulf- and Atlantic coastal plains [Rawls and Turnquist, 2010]

It is assumed that links as well as facilities - and hence the stored relief items - can be destroyed partially or completely by a disaster such that demand may not be sufficiently satisfied. Uncertainties like unknown demand and damage levels are captured through the use of a finite number of disaster scenarios with corresponding probabilities of occurrence. In particular, each scenario contains the information on demand, available road capacities and the proportion of undamaged relief items depending on the potential disaster outcome.

The objective is to minimize the overall costs, consisting of costs for opening facilities and storing aid supplies as well as expected costs for transporting relief items, penalty costs for unsatisfied demand and unused (leftover) items. In the latter case costs may occur for the disposal of, e.g., expired medicine.

Before presenting the two-stage stochastic Rawls and Turnquist [2010] model, the notation is given below. First-stage parameters are known at the time of preparation, whereas second-stage parameters depend on a particular scenario and are therefore uncertain:

Sets:

I: set of demand nodes

K: set of relief items

L: set of facility size categories

N: set of nodes, $I \subseteq N$

R: set of arcs in the network, $R \subseteq N \times N$

S: set of possible scenarios

Scenario-independent parameters:

b^k: space requirement per unit for relief item $k \in \mathrm{K}$

$\mathrm{c}^k_{(i,j)}$: costs for transporting one unit of relief item $k \in \mathrm{K}$ on link $(i,j) \in R$

F_l: costs for opening facility of size $l \in \mathrm{L}$

h^k: penalty cost per unit for unused relief item $k \in \mathrm{K}$, e.g. holding or disposal costs

M_l: capacity of a facility of size $l \in \mathrm{L}$

p^k: penalty cost per unit for unsatisfied demand of relief item $k \in \mathrm{K}$

q^k: acquisition cost per unit for relief item $k \in \mathrm{K}$

u^k: required transport capacity per unit of relief item $k \in \mathrm{K}$

Scenario-dependent parameters:

P_s: probability of scenario $s \in \mathrm{S}$

$\rho_i^{k\,s}$: proportion of relief item $k \in \mathrm{K}$ at node $i \in \mathrm{I}$ in scenario $s \in \mathrm{S}$ which is still usable

$\mathrm{U}^s_{(i,j)}$: maximal available arc capacity of link $(i, j) \in \mathrm{R}$ in scenario $s \in \mathrm{S}$

$\mathrm{v}_i^{k\,s}$: demand for relief item $k \in \mathrm{K}$ at node $i \in \mathrm{I}$ in scenario $s \in \mathrm{S}$

Scenario-independent decision variables (first stage):

r_i^k: pre-positioned quantity of relief item $k \in \mathrm{K}$ at node $i \in \mathrm{I}$

y_{il}: 1 if facility of size $l \in \mathrm{L}$ is opened at node $i \in \mathrm{I}$; 0 otherwise

Scenario-dependent decision variables (second stage):

$w_i^{k\,s}$: unsatisfied demand for relief item $k \in \mathrm{K}$ at node $i \in \mathrm{I}$ in scenario $s \in \mathrm{S}$

$x^{k\,s}_{(i,j)}$: quantity of relief item $k \in \mathrm{K}$ transported from i to j in scenario $s \in \mathrm{S}$

$z_i^{k\,s}$: unused quantity of relief item $k \in \mathrm{K}$ at node $i \in \mathrm{I}$ in scenario $s \in \mathrm{S}$

The model formulation is as follows, where the objective is to minimize over the first-stage variables y_{il} and r_i^k, and second-stage variables $x_{(i,j)}^{ks}$, z_i^{ks} and w_i^{ks}, given the parameters F_l, q^k, $\mathrm{c}_{(i,j)}^k$, h^k and p^k:

$$\begin{aligned} \min \sum_{i \in \mathrm{I}} \sum_{l \in \mathrm{L}} \mathrm{F}_l y_{il} + \sum_{k \in \mathrm{K}} \sum_{i \in \mathrm{I}} \mathrm{q}^k r_i^k \\ + \sum_{s \in \mathrm{S}} P_s \left[\sum_{(i,j) \in \mathrm{R}} \sum_{k \in \mathrm{K}} \mathrm{c}_{(i,j)}^k x_{(i,j)}^{ks} + \sum_{i \in \mathrm{I}} \sum_{k \in \mathrm{K}} (\mathrm{h}^k z_i^{ks} + \mathrm{p}^k w_i^{ks}) \right] \end{aligned} \tag{2.1}$$

s.t.
1. One facility per node

$$\sum_{l \in L} y_{il} \leq 1 \quad \forall i \in \mathrm{I} \tag{2.2}$$

2. Facility capacity

$$\sum_{k \in K} \mathrm{b}^k r_i^k \leq \sum_{l \in L} \mathrm{M}_l y_{il} \quad \forall i \in \mathrm{I} \tag{2.3}$$

3. Arc capacity

$$\sum_{k \in \mathrm{K}} \mathrm{u}^k x_{(i,j)}^{ks} \leq \mathrm{U}_{(i,j)}^s, \quad \forall (i,j) \in \mathrm{R}, s \in \mathrm{S} \tag{2.4}$$

4. Flow conservation

$$\sum_{j \neq i \in \mathrm{N}} x_{(j,i)}^{ks} + \rho_i^{ks} r_i^k - z_i^{ks} = \sum_{j \neq i \in \mathrm{N}} x_{(i,j)}^{ks} + \mathrm{v}_i^{ks} - w_i^{ks}, \quad \forall i \in \mathrm{N}, k \in \mathrm{K}, s \in \mathrm{S} \tag{2.5}$$

5. Binary and non-negative decision variables

$$y_{il} \in \{0, 1\} \quad \forall i \in \mathrm{I}, l \in \mathrm{L}, \tag{2.6}$$

$$r_i^k \geq 0 \quad \forall i \in \mathrm{I}, k \in \mathrm{K}, \tag{2.7}$$

$$x_{(i,j)}^{ks} \geq 0, \quad \forall (i,j) \in \mathrm{R}, k \in \mathrm{K}, s \in \mathrm{S}, \tag{2.8}$$

$$z_i^{ks}, w_i^{ks} \geq 0, \quad \forall i \in \mathrm{I}, k \in \mathrm{K}, s \in \mathrm{S}. \tag{2.9}$$

The objective (2.1) is to minimize the first-stage costs (first line), i.e. fixed costs for opening facilities of specific size and pre-positioning costs, and the expected second-stage costs (second line) composed of transportation costs as well as penalty costs for unused items and unsatisfied demand. Note that only first-stage costs are actually incurred when location and inventory decisions are carried out prior to the event of a disaster. In contrast, costs associated with second-stage decisions represent just auxiliary terms. For instance, transportation costs arise only in the aftermath of a disaster when relief items are actually delivered to the affected population. Moreover, penalty cost rates h^k and p^k are used to avoid either excessive storage and unsatisfied demand, respectively. Especially penalty costs concerning unsatisfied demand are purely fictitious but imply a higher demand satisfaction on the second stage.

According to (2.2), no more than one facility per node is permitted to be opened. If a facility of size l is opened at node i, i.e. $y_{il} = 1$, the quantity of pre-positioned items cannot exceed the corresponding capacity M_l; see (2.3). Both constraint groups (2.2) and (2.3) belong to the first stage with scenario-independent parameters b^k and M_l. However, facility and pre-positioning decisions, y_{il} and r_i^k, directly influence second-stage decisions concerning, e.g., transport quantities $x^{k\,s}_{(i,j)}$.

According to (2.4), it is not allowed to transport more items along an arc than the available arc capacity $U^s_{(i,j)}$. In case of road destruction, where $U^s_{(i,j)} = 0$ in (2.4) for a specific scenario s, delivering relief items is impossible. The possibility of damaged road network conditions and the related risk has to be considered when location decisions are made on the first stage.

The flow conservation constraints in (2.5) ensure that all goods available in a node are either used to satisfy demand, transported to another node or stay at that node as a surplus, for every node i, relief item k and scenario s. Leaving demand unsatisfied is also possible. Depending on the magnitude and the location of the disaster, a proportion $\rho_i^{k,s} \in [0, 1]$ of relief items remains intact after the disaster and can be used to meet demand. In the case of $\rho_i^{k,s} = 0$, items of type k pre-positioned at node i are completely destroyed. Here, the second stage is linked to the first stage by the first-stage decision variables r_i^k. In other words, second-stage variables like unsatisfied demand $w_i^{k\,s}$ depend on where and how many relief items have been pre-positioned on the first stage. Hence, constraints (2.5) are also referred to as *linking constraints*. It can be expected that if shortages are penalized more than excess inventory, relief items are rather pre-positioned at safer facilities, i.e. those with higher survival rate $\rho_i^{k,s}$.

All variables should be non-negative and variables $y_{i,l}$ are binary as defined in (2.6) - (2.9).

Note that the first-stage decisions, namely the location and size of the facilities $y_{i,l}$ and the amounts of pre-positioned relief items r_i^k, are the main results of the model. In contrast, the second stage is primarily used to support appropriate first-stage decisions by anticipating possible future developments, and not for developing concrete delivery strategies. Hence, only a network flow problem is formulated on the second stage. After the realization of a specific disaster scenario, a vehicle routing problem can be solved in practice to obtain a better and more detailed plan for the last mile distribution of aid supply.

2.3.2 Extensions

After the publication of Rawls and Turnquist [2010] several authors extended or modified the benchmark model (2.1)-(2.9) by covering additional aspects or by compensating its weaknesses. For instance, Hong et al. [2015] overcome the estimation of penalty costs for unsatisfied demand p^k by introducing chance constraints, where specific constraints have to be fulfilled with a predefined probability α. In particular, the authors require that the uncertain demand v_i^{ks} in (2.5) has to be satisfied with a probability α, which is the required reliability that has to be specified by the user:

$$P\left(\sum_{j\neq i\in N} x_{(j,\,i)}^{k,\,s} + r_i^k - \sum_{j\neq i\in N} x_{(i,\,j)}^{k,\,s} \geqslant v_i^{k,\,s}\right) \geqslant \alpha \quad \forall i \in N,\ k\in K,\ s\in S.$$

Such formulations allow Hong et al. [2015] to avoid the explicit determination of penalty costs, which is very difficult in a humanitarian context. Numerical results show that their model is superior to the one proposed by Rawls and Turnquist [2010] with respect to demand satisfaction.

Another extension of the Rawls and Turnquist [2010] model is proposed by Noyan [2012], where the cVaR is incorporated as a risk measure. In this case the objective function (3.1) is replaced by a risk-averse version:

$$\min\ (1+\lambda)\cdot a^T\chi + E_\xi Q(\chi,\xi(\omega)) + \lambda\cdot(Q(\chi,\xi(\omega)))\cdot \text{cVaR}, \tag{2.10}$$

where $\lambda \geq 0$ is a risk parameter. With higher values of λ, the risk term is weighted more in relation to the expected second-stage costs, reflecting the risk averseness of the decision maker. As a result, overall more relief items will be stored than in the risk-neutral approach, where only the expected second-stage costs are taken into account. As stated in Noyan [2012], solutions found by the risk-averse approach are therefore more robust with respect to the realization of extreme disaster events than solutions generated by the model of Rawls and Turnquist [2010].

Rawls and Turnquist [2011] extend their own model (2.1)–(2.9) by defining a so-called reliable set consisting of scenarios where unmet demand is not allowed, i.e. $w_i^{ks} \neq 0$. To ensure storage of critical commodities near the disaster location, additional *service quality* constraints are proposed of the form:

$$\sum_{(i,j)\in R} d_{(i,j)} x_{(i,j)}^{ks} \leq D^k \sum_{i\in N} v_i^{ks} + O\left(1-\psi^s\right) \qquad \forall k\in K,\ s\in S,$$

where $d_{(i,j)}$ denotes the distance between nodes i and j, D^k is the acceptable distance limit for item k, O is a large number and the binary variable ψ^s takes value 1 if scenario s is part of the reliable set. For $\psi^s = 1$ the distances over all arcs in the network weighted by the transportation quantity, $\sum_{(i,j)\in R} d_{(i,j)} x_{(i,j)}^{k,s}$, cannot exceed the maximal allowed demand weighted distance $D^k \sum_{i\in N} v_i^{k,s}$ for relief item k, i.e. "on average", the distance limit D^k may not be exceeded. As a result, facilities for more critical items are opened closer to demand points. Since D^k is chosen depending on the level of urgency, a more differentiated handling of aid supplies is possible than in the original model.

3 Solution Algorithms in Disaster Management

Quantitative optimization models as those reviewed in the previous chapter are of vital importance for the decision-making process in disaster management. However, solving such models within a reasonable time frame is a necessary but challenging task. In particular, two-stage stochastic programs become very large for practical applications slowing down the pace of solution processes considerably. As identified in the previous chapter, relief operations are carried out under a high time pressure requiring fast solutions. For instance, some natural disasters like storms and floods can be predicted up to five days in advance. During these days aid organizations can initiate evacuations and pre-position relief items at convenient locations. Time for solving the relevant two-stage stochastic model should fit into the tight schedule of such preparation strategies. Only if efficient methods providing fast solutions exist, two-stage stochastic models are applicable in humanitarian practice. Therefore, many of the publications summarized in Table 2.1 propose, in addition to the model formulation, solution algorithms. These are discussed in the following section and can be roughly divided into exact methods (Section 3.1.1) and heuristics (Section 3.1.2). However, both approaches have severe disadvantages. One main drawback of heuristics is that solution quality in general is not measurable as heuristics are often based on rules of thumb. Especially in cases of emergency, optimality of the solution can be a matter of life and death. On the other side, exact solution algorithms like the simplex algorithm or interior-point methods are too time-consuming for large-scale problems arising in realistic settings. Therefore, a method should be developed which provides a near-optimal solution within an acceptable computational time.

The so-called L-shaped method [Slyke and Wets, 1969] is particularly suited for solving two-stage stochastic programs and belongs to the class of exact methods. Therefore, this is the method which is used and developed further in this work.

After introducing the theoretical background of two-stage stochastic programming in Section 3.2.1 the classical version of the L-shaped method[1] is described in Section 3.2.2. In order to solve more realistic large-scale two-stage stochastic problems the L-shaped method has to be enhanced further. The basic idea of how the L-shaped method can be accelerated is motivated in Section 3.3.1. Afterwards, Section 3.3.2 indicates what

[1]Note that the standard L-shaped method is applicable for continuous and/or integer first-stage variables and continuous second-stage variables. In the case of integer second-stage variables the integer version of the L-shaped method has to be used [Laporte and Louveaux, 1993].

E. Graß, *An Accelerated Solution Method for Two-Stage Stochastic Models in Disaster Management*, Mathematische Optimierung und Wirtschaftsmathematik | Mathematical Optimization and Economathematics,
https://doi.org/10.1007/978-3-658-24081-3_3

requirements have to be met by two-stage stochastic programs such that the accelerated L-shaped method can be applied. The mathematical details are given in Section 3.3.3.

3.1 Solution Methods in Disaster Management: A Literature Review

Table 3.1[2] shows solution methods which are used by the different authors for solving two-stage stochastic programs in disaster management.[3] Before implementing and solving the proposed models, some modifications are required in certain cases. For instance, the so-called defuzzification techniques allow a so-called *deterministic crisp counterpart* to be formulated to the respective possibilistic model [Tofighi et al., 2016]. Another example is that multiple objectives have to be converted into a single-objective by using, e.g., goal programming [Bai, 2016], compromise programming [Bozorgi-Amiri et al., 2013] or the ϵ-constraint approach [Rezaei-Malek et al., 2016, Tricoire et al., 2012]. The ϵ-constraint method is often used to convert a multi-objective problem into its single-objective counterpart [Ehrgott, 2005, Ch. 4.1]. Only one objective is chosen as the main objective function whereas the remaining objectives are formulated as constraints bounded by an ϵ-value.

In all the described cases, only approximate solutions for the original problem will result. However, when the reformulated problems are solved by a commercial solver, the corresponding solution is stated as exact in Table 3.1, as an exact method is applied.

Table 3.1: Solution methods

Author(s)	Year	Solution Method	Solution	
			exact	non-exact
Barbarosoglu and Arda	2004	commercial solver	x	
Chang et al.	2007	commercial solver	x	
Balcik and Beamon	2008	commercial solver	x	
Andreas and Smith	2009	L-shaped method with valid inequalities	x	
Liu and Huang	2009	transformation into two sub-problems; commercial solver	x	
Liu et al.	2009	L-shaped method	x	
Ng and Waller	2009	commercial solver	x	
Fan and Liu	2010	progressive hedging (PH) heuristic		x
Mete and Zabinsky	2010	commercial solver	x	
Peeta et al.	2010	first-order approximation		x
Rawls and Turnquist	2010	Lagrangian L-shaped method with greedy heuristics		x
Salmeron and Apte	2010	commercial solver	x	
Cai et al.	2011	transformation into two sub-problems; commercial solver	x	
Falasca and Zobel	2011	commercial solver	x	

(continued on the next page)

[2]The table is taken from Grass and Fischer [2016b].

[3]This section is partially taken from Grass and Fischer [2016b].

Table 3.1: Solution methods (continued)

Author(s)	Year	Solution Method	Solution exact	non-exact
Li et al.	2011	L-shaped method	x	
Rawls and Turnquist	2011	commercial solver	x	
Bozorgi-Amiri et al.	2012	particle swarm optimization (PSO) meta-heuristic		x
Döyen et al.	2012	Lagrangian method with local search heuristic		x
Li et al.	2012	heuristic with Lagrangian method		x
Lodree et al.	2012	percentage of demand scenarios solution approach (PDSA)		x
Noyan	2012	L-shaped method with multi-cuts	x	
Rawls and Turnquist	2012	commercial solver	x	
Tricoire et al.	2012	exact and heuristic algorithm	x	x
Wang et al.	2012	commercial solver	x	
Bozorgi-Amiri et al.	2013	commercial solver	x	
Davis et al.	2013	commercial solver	x	
Klibi et al.	2013	commercial solver	x	
Maqsood and Huang	2013	transformation into two sub-problems; commercial solver	x	
Wang and Huang	2013	transforming the flexible fuzzy decision space into four fixed scenarios; commercial solver	x	
Ahmadi et al.	2015	variable neighborhood search heuristic		x
Hong et al.	2015	commercial solver	x	
Hu et al.	2015	commercial solver	x	
Verma and Gaukler	2015	L-shaped Method with greedy and local search heuristic	x	
Zolfaghari and Peyghaleh	2015	commercial solver	x	
Alem et al.	2016	two-phase heuristic		x
Bai	2016	commercial solver	x	
Noyan et al.	2016	integer L-shaped Method with mutli-cuts, lazy constraints (CPLEX) and parallelization	x	
Rath et al.	2016	commercial solver	x	
Rezaei-Malek et al.	2016	separable programming; commercial solver	x	
Tofighi et al.	2016	differential evolution heuristic		x

3.1.1 Exact Methods

Table 3.1 reveals that a considerable number of authors choose a commercial solver like CPLEX[4] for solving their respective model directly. However, the software cannot exploit the two-stage structure, resulting in time- and memory-consuming calculations. Therefore, more suitable exact and heuristic procedures are proposed in the literature and discussed in the following.[5]

[4]More information on the IBM ILOG CPLEX Optimizer can be found on https://www.ibm.com/us–en/marketplace/ibm–ilog–cplex.

[5]As this section describes different approaches proposed to solve two-stage stochastic programs, contributions where only a commercial solver is applied are not reviewed here.

As shown in Table 3.1, some authors use the L-shaped method [Slyke and Wets, 1969] to solve their two-stage model to optimality [Li et al., 2011]. The basic idea of this method is to solve the first-stage problem, called *master problem*, and the second-stage problem, consisting of several *sub-problems*, separately. In particular, the master problem and the sub-problems are solved successively again and again until the optimal solution or a sufficiently accurate approximation of the original problem is obtained (see Section 3.2.2 for details).

In general, the L-shaped method outperforms direct approaches like the Branch-and-Bound or Branch-and-Cut algorithm as it solves several smaller problems instead of one large-scale problem. However, it may happen that the L-shaped method stagnates at the end of the process, converging only slowly to the optimal solution [Rahmaniani et al., 2017]. For instance, the case study of Li et al. [2011] is of rather moderate size, containing 45 scenarios and 7,695 decision variables, of which 57 are binary, but the L-shaped method can find the solution only after 206 iterations and 3,509 seconds (about one hour). Computational time for solving the model by a commercial solver is not provided.

In order to accelerate the L-shaped method, Andreas and Smith [2009] add two types of inequalities to the master problem. The first type of inequalities ensures the feasibility of second-stage solutions and the second type of inequalities provides information on the lower bound of the second-stage objective function value to the master problem.

In general, the solution of the master problem becomes computationally expensive when integer or binary first-stage variables are present. Especially in the pre-positioning setting location decisions are modeled by binary variables. In order to avoid solving the master problem exactly numerous times, Verma and Gaukler [2015] combine the greedy heuristic with a local search for the calculation of the first-stage decisions. In particular, the master problem is solved heuristically until two successively computed first-stage solutions are identical. In this case the heuristic does not lead to further improvement and CPLEX is used to solve the master problem exactly, providing an updated first-stage solution. This procedure is terminated when the overall solution is obtained. The computational experiments in Verma and Gaukler [2015] show a significant acceleration of solution time for their heuristically modified L-shaped method in contrast to solving the extensive form via CPLEX.

3.1.2 Heuristics

For large-scale realistic problems, heuristics may represent the only way to find solutions within a reasonable time frame. The corresponding first-stage decisions may even be better than the ones derived by exact methods from a simplified problem formulation using only a small number of scenarios.

Table 3.2[6] provides a classification of the different heuristic approaches suggested for two-stage stochastic programs, namely heuristics which are tailored to a specific problem,

[6] The table is taken from Grass and Fischer [2016b].

metaheuristics and matheuristics. The latter combine (meta-)heuristics with techniques taken from mathematical programming.[7] In general, heuristics are especially valuable for large-scale problems where exact solution techniques cannot be used. Besides their advantages, also some drawbacks and limitations of specific heuristics are pointed out below. For instance, the quality of a heuristic solution generally cannot be judged a priori. Therefore, the authors who are cited here use exact solutions or bounds as a basis for a comparison, allowing the evaluation of their respective heuristic approach. One exception is the publication of Li et al. [2012], where no assessment of the solution quality is given, making it difficult to evaluate the proposed heuristic. However, in this case the calculation of an exact solution in closed form is not straightforward due to their complex bi-level problem structure.

Table 3.2: Types of heuristics

Author(s)	Year	Problem-tailored heuristic	Metaheuristic	Matheuristic
Fan and Liu	2010			x
Peeta et al.	2010	x		
Rawls and Turnquist	2010			x
Bozorgi-Amiri et al.	2012		x	
Döyen et al.	2012			x
Li et al.	2012			x
Lodree et al.	2012	x		
Tricoire et al.	2012	x		
Ahmadi et al.	2015		x	
Alem et al.	2016	x		
Tofighi et al.	2016		x	

Matheuristics

Rawls and Turnquist [2010] propose the so-called Lagrangian L-shaped method (LLSM), where constraints (2.3) are included in the objective function (2.1) of the master problem using Lagrange multipliers.[8] These multipliers are updated via a sub-gradient method, resulting in a lower bound for the objective function (2.1). By relaxing constraints (2.3), the master problem can be divided into two problems, one problem for determining pre-positioning decision variables r_i^k and one location problem with decision variables y_{il}. The latter problem is solved via the greedy heuristic, such that optimality of the overall solution cannot be guaranteed. In order to reduce the complexity of the second-stage sub-problems, Rawls and Turnquist [2010] either assume link capacities have the value 0 for certain scenarios, i.e. $U_{ij}^s = 0$, such that no transport can be carried out on damaged

[7] For a detailed description of matheuristics, see Maniezzo et al. [2010].

[8] The classical Lagrange method is used to integrate equality constraints into the objective function by means of Lagrange multipliers; see for example Arora [2015, Ch.6] for details.

links, or assume that transportation quantities are unlimited for undamaged links. As a result, the arc capacity constraints (2.4) become obsolete and the multi-commodity network flow is replaced by a series of single-commodity network flows. For every type of relief item a single-commodity network flow model is solved, leading to a simplification of the second-stage problems; however, this simplification also limits the applicability of this approach. The LLSM requires 51 seconds to find a near-optimal solution for a case study consisting of 147,060 decision variables (including 90 binary variables) and 7,608 constraints. In particular, the objective function value differs 0.6% from the exact solution, which is obtained after around 12 minutes by CPLEX.

The following heuristics do not decompose the problem into a master problem and sub-problems as the L-shaped method. However, Döyen et al. [2012] relax the flow conservation constraints in their model, similar to Rawls and Turnquist [2010], via the Lagrangian method, allowing the separation of the original problem into two smaller sub-problems. While the first sub-problem contains location and pre-positioning decisions and is solved by CPLEX, the second sub-problem contains decisions concerning shortages and transportation quantities and can be subdivided further, again applying the Lagrangian method. Then the local search heuristic is used, which determines suitable neighborhood solutions in every iteration until a predefined time level is reached. The numerical results of Döyen et al. [2012] indicate that their solution approach provides feasible solutions within two hours which are on average 2.55% away from the best available lower bound. In contrast, CPLEX computes a solution with a deviation of 11.58% from the best lower bound on average within the same time frame. A drawback of the local search heuristic is that it can find only a local optimum. To avoid such a situation, the tabu search heuristic can be applied, where a list is created with solutions that have already been evaluated and should not be generated again for a certain number of iterations, and where worse solutions can be accepted, allowing the escape from a local minimum [Glover and Laguna, 1997]. However, none of the publications reviewed here uses this technique. The reason might be that the user has to define in which case a solution should be put onto the tabu list requiring good knowledge of the underlying problem [Shen et al., 2000].

Li et al. [2012] relax shelter capacity constraints of their two-stage stochastic program by formulating the Lagrangian function. Due to the concavity of the Lagrangian dual problem, they can apply a line search heuristic where the search directions are based on calculated sub-gradients. The two-stage stochastic program represents one level of the bi-level model proposed by Li et al. [2012]; see Section 2.2.1. Hence, this heuristic is embedded into another heuristic intended to solve the main bi-level problem. As mentioned above, neither solution quality nor solution time regarding the proposed heuristic are given.

Before describing the approach of Fan and Liu [2010] the definition of convex problems has to be introduced. Convex problems consist of constraints and an objective which are convex functions. If the domain of a function $f : \mathbb{R}^n \to \mathbb{R}$ is a convex set and for all x_1, x_2 in this domain and all $\varrho \in [0, 1]$ the inequality $f((1-\varrho)x_1 + \varrho x_2) \leq (1-\varrho)f(x_1) + \varrho f(x_2)$ is fulfilled, then f is called convex [Boyd and Vandenberghe, 2004, p. 67]. For non-convex functions this inequality does not hold. Since non-convex problems cannot be solved by the L-shaped method [Slyke and Wets, 1969], Fan and Liu [2010] apply the *progressive hedging* technique as a heuristic[9] for their non-convex and discrete model. The progressive hedging method decomposes the underlying problem into several network design problems of smaller size. In order to further simplify these sub-problems, relaxation techniques are used leading to non-linear programs, which are solvable by using a commercial solver like GAMS[10]. After obtaining the corresponding solutions, the augmented Lagrangian has to be solved in every iteration of the progressive hedging method. Although the numerical example introduced by Fan and Liu [2010] is relatively small (27 nodes, 76 links and 64 scenarios), a solution time of about two hours is reported. No computational time for solving the model by GAMS is provided.

Metaheuristics

The advantage of metaheuristics is that they are applicable to many problems. However, they may suffer from the disadvantage of stagnating in local optima. Hence, Ahmadi et al. [2015] use the *variable neighborhood search* heuristic for the multi-depot location-routing problem to escape such local minima. During the solution procedure, the neighborhood structure is changed according to the so-called shaking phase. For instance, the positions of demand nodes are swapped randomly within a tour so that no adjustments of parameters on the part of the decision maker are required. This heuristic leads to around 16% less satisfied demand than the exact solution for the model proposed by Ahmadi et al. [2015]. For major disasters such inaccurate solutions may have devastating consequences for those affected.

Tofighi et al. [2016] transform their fuzzy-based multi-objective model into an equivalent deterministic single objective problem in several reformulation steps. Only then the *differential evolution* heuristic can be applied. At every iteration of this heuristic, new solutions are added to the present population by means of mutation and crossover of potential solutions. If the maximum number of iterations is reached or a sufficient improvement of the solution cannot be observed, the evolution process is terminated. In order to ensure a solution pertaining to the particular requirements of practitioners, the decision maker, e.g. coordinator of a relief organization, is integrated in the entire solution process by choosing the weights for the objectives or other control parameters. It is a disadvantage of the differential evolution heuristic that its performance depends on the parameter selection which is based on a trial-and-error approach [Tofighi et al., 2016]. As stated by the same authors, the computation time of the differential evolution heuristic exceeds four hours for a realistic case study with more than 291,940 integer variables. Solution times for a commercial solver are not reported. A relative optimality gap[11] of around 1.89% on average is given only for small size examples, but not for the proposed case study.

[9]Convergence of the standard progressive hedging algorithm is guaranteed for continuous problems only [Rockafellar and Wets, 1991].

[10]More information on the GAMS optimization system can be found on https://www.gams.com/.

[11]The relative optimality gap is defined as $(OV_H - OV_E)/(OV_E)$, where OV_H denotes the objective value of the heuristic and OV_E of the exact method.

Another population-based meta-heuristic is designed by Bozorgi-Amiri et al. [2012], where the *particle swarm* heuristic is adjusted to problems with continuous and binary variables. As described above, finding only local optima has to be avoided. Hence, after a predefined number of iterations the solution method selects and re-initializes some particles randomly so that candidate solutions are varied periodically. Their solution procedure achieves an average deviation of 4.1% from the optimal value for small- and medium-scale problems. Since optimal solutions cannot be obtained for large-scale instances within 30 minutes, Bozorgi-Amiri et al. [2012] compare the heuristically computed solutions after 100, 300 and 600 iterations. The results show that the objective function values obtained after 100 and 600 iterations differ only slightly. However, as the problem size increases to 40 scenarios, 25 suppliers, 35 potential relief distribution centers and 35 affected areas, the difference between these objective function values increases.

Problem-Tailored Heuristics

Some authors suggest heuristics which are especially tailored to the specific model structure they consider. Peeta et al. [2010] reformulate their two-stage stochastic program as a knapsack problem, the solution of which is the local minimum of the original problem. Numerical results published by Peeta et al. [2010] show only minor differences between the local and global minimum, so an escape from a local optimum is not required. The idea of problem reformulation is also used in Lodree et al. [2012]. They call their solution procedure the *Percentage of Demand Scenarios solution Approach* (PDSA), which is split into two steps. First, the quantities to be pre-positioned on the first stage are calculated based on two cases, the conservative and the aggressive pre-positioning strategy. The conservative strategy is pursued when the expected overstock for scenarios *without* demand is more expensive than the expected shortage for scenarios *with* demand. In the opposite case, the aggressive strategy is preferred. As a result, the inventory level is either limited to the minimum demand in the conservative case or is at least the minimum demand in the aggressive case. After setting the pre-positioning quantities, a transportation problem is solved in the second step via the Excel Solver^{TM}[12]. The numerical results of Lodree et al. [2012] show that the solution found by PDSA depends on the penalty costs for unmet demand as well as on the distance between the manufacturer and the retailers. Since the manufacturer sends relief items to the retailer where they are stored before the occurrence of a hurricane, the corresponding distance influences the pre-positioning decisions. In particular, PDSA performs the worse, the further away the manufacturer is located from the retailers and/or the lower the shortage costs are. As stated in Lodree et al. [2012], solutions produced by the PDSA are on average 3.6% above the optimal objective value. No CPU times are reported.

A two-phase heuristic is proposed by Alem et al. [2016], where the underlying model is simplified such that the number of integer decision variables is reduced. The first phase is concerned with solving the simplified problem providing transported quantities (second-

[12]More information on Excel Solver^{TM} can be found on http://opensolver.org/.

stage decisions). Afterwards, in the second phase, the original problem is solved with these quantities as input parameters. The numerical results indicate that the two-phase heuristic requires 500 seconds (less than 10 minutes) to find a solution for the minimax-regret model in the worst case whereas CPLEX terminates after 14,400 seconds (about four hours) at the most. However, the corresponding solutions have on average a relative optimality gap of 20.07% (and 122.9% in the worst case). The remaining relative gaps are as follows: 0.3613% on average for the risk-neutral two-stage stochastic program, 0.0076% for the version with cVaR and 0.1957% with the semi-deviation measure. Hence, their heuristic performs best for the cVaR model and is not recommended for the minimax-regret model.

Tricoire et al. [2012] choose the ϵ-constraint approach for solving their bi-objective model. Here, the objective function concerning the total costs is minimized where additional ϵ-constraints restrict the second objective function, namely the expected uncovered demand. Afterwards, the corresponding solutions are used to minimize the second objective function leading to a Pareto-optimal solution.[13] In every iteration the ϵ-constraints are updated providing a set of Pareto-optimal solutions, the Pareto front. In order to find at least an approximated Pareto front for large-scale examples, Tricoire et al. [2012] gradually increase the optimality gap of the commercial solver and the allowable time limit. Numerical tests show a maximum optimality gap of 70%, highlighting the shortcomings of such a heuristic approach.

3.2 Two-Stage Stochastic Programming

The next section describes the theoretical principles of two-stage stochastic programming. The question how to solve such programs via the L-shaped method is addressed in Section 3.2.2. With these both sections the foundations for the development of an improved solution algorithm, namely the accelerated L-shaped method in Section 3.3, are laid.

3.2.1 Introduction

As already outlined in Section 2.2.1, the set of decision variables in two-stage stochastic models is split into two parts, namely decisions made before and after the realization of an uncertain event like the occurrence of a disaster.[14] First-stage decisions, denoted by the vector $\chi \in \mathbb{R}^{n_1}$, are taken under uncertainty when complete information is not yet available. After the uncertainty has been unfolded, some recourse action can be taken depending on the respective outcome of the event. For instance, in preparation for a disaster, relief items can be stored at specific locations on the first stage before an emergency situation occurs. Depending on the magnitude and the location of the actual disaster, a possible recourse action of the relief organization could be to buy additional commodities to satisfy additional demand [Falasca and Zobel, 2011].

[13] A solution is denoted as Pareto-optimal if it is not possible to improve it for one objective without worsening the other objective.

[14] The following text (to page 46) is partially taken from Grass and Fischer [2016b].

In general, a two-stage stochastic linear program with recourse can be stated as[15]

$$\min\ a^T\chi + E_\xi\left[Q(\chi, \xi(\omega))\right] \tag{3.1}$$

$$\text{s.t. } G\chi = d \tag{3.2}$$

$$\chi \geq 0, \tag{3.3}$$

where $a \in \mathbb{R}^{n_1}$, $d \in \mathbb{R}^{m_1}$ are the deterministic first-stage vectors, $G \in \mathbb{R}^{m_1 \times n_1}$ is the corresponding constraint matrix and $\xi(\omega)$ is a random vector containing uncertain data which depend on the random event ω. In the humanitarian context, ω can be interpreted as the occurrence of a disaster. The *recourse function* $Q(\chi, \xi(\omega))$ of the linear second-stage problem is as follows:

$$Q(\chi, \xi(\omega)) = \min\{f^T\gamma_\omega \mid T(\omega)\chi + W(\omega)\gamma_\omega = t(\omega), \gamma_\omega \geq 0\}, \tag{3.4}$$

where the objective is to minimize the second-stage costs $f^T\gamma_\omega$ incurred by the second-stage decisions $\gamma_\omega \in \mathbb{R}^{n_2}$ for each ω. The objective function in (3.1) is to minimize the costs associated with the first-stage decisions χ, namely $a^T\chi$, plus the expected second-stage costs. The expectation is denoted by E, depending on ξ. In particular, the random vector $\xi(\omega) = (t(\omega), T(\omega), W(\omega))$ contains uncertain second-stage parameters. Here, the notation (ω) indicates the functional dependence of the parameters on ω whereas γ_ω denotes the solution of the linear second-stage problem for each ω. The so-called *recourse matrix* $W(\omega) \in \mathbb{R}^{m_2 \times n_2}$, *technology matrix* $T(\omega) \in \mathbb{R}^{m_2 \times n_1}$, as well as the right-hand side $t(\omega) \in \mathbb{R}^{m_2}$ vary with the realization of a random event ω. The optimal value $Q(\chi, \xi(\omega))$, i.e. the minimal second-stage costs for each random event ω, has to be found such that the corresponding constraints

$$T(\omega)\chi + W(\omega)\gamma_\omega = t(\omega), \quad \gamma_\omega \geq 0,$$

are fulfilled. The recourse is called *fixed* if W does not depend on ω and it is called *complete* if for every first-stage decision there exists γ_ω which satisfies the second-stage constraints. For instance, the Rawls and Turnquist [2010] model has a fixed and complete recourse. In particular, for any first-stage decisions y_{il} and r_i^k there exist feasible second-stage decisions w_i^{ks}, $x_{(i,j)}^{ks}$ and z_i^{ks}. In other words, regardless of the number of opened facilities and the quantity of pre-positioned relief items, second-stage variables can take feasible values which result either in unsatisfied demand w_i^{ks} or in a surplus of relief items z_i^{ks}. Hence, second-stage constraints (2.4) and (2.5) are always fulfilled.

In the following, it is assumed that ω leads to one scenario from a previously defined set of scenarios S, i.e. the general notation ω is replaced by s with $s \in S$.[16] Every scenario occurs with a specific probability P_s such that $\sum_{s \in S} P_s = 1$ and the number of scenarios is finite. The expected recourse function $E_\xi\left[Q(\chi, \xi(s))\right]$ in (3.1) results from the sum of

[15]The following presentation is based on Birge and Louveaux [2011].

[16]Note that Birge and Louveaux [2011, p. 21] define ω in a more general sense, namely as random events, whereas scenarios denote a set of the outcomes sharing common properties. For instance, each scenario may contain the information on demand, i.e. demand represents the common property.

$Q(\chi, \xi(s))$ weighted by the scenario probability P_s. The recourse function $Q(\chi, \xi(s))$ is piece-wise linear, as it results from a finite number of linear second-stage constraints, and is convex in χ.[17] Since the sum of convex and piece-wise linear functions is again convex and piece-wise linear, so is the corresponding expected recourse function $E_\xi\left[Q(\chi, \xi(s))\right]$ [Higle, 2006]. Such a function is shown in Figure 3.1, where the expected second-stage costs depend on the first-stage decisions χ. For instance, χ^2 leads to higher expected second-stage costs than χ^1.

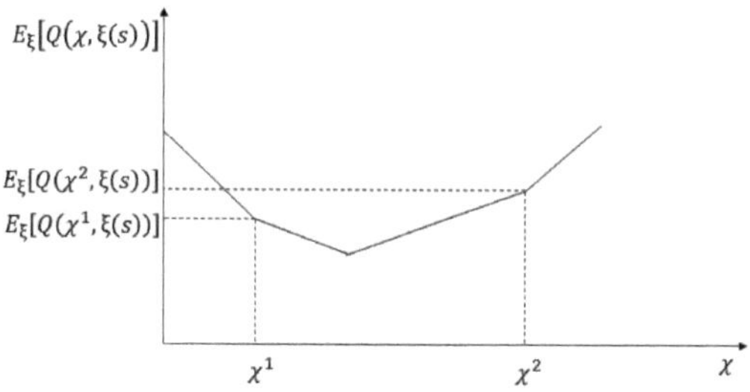

Figure 3.1: Example of a piece-wise linear expected recourse function $E_\xi\left[Q(\chi, \xi(s))\right]$

In order to get a better idea of the respective matrix structures, and to relate the Rawls and Turnquist [2010] model (2.1)-(2.9) to the general formulation in (3.1)-(3.4) a small example is given below[18]:

Example

Assume a problem with three scenarios $|S| = 3$ and three nodes $|N| = |I| = 3$, i.e. $|R| = 9$. In addition, two types of relief items $|K| = 2$ and facility sizes $|L| = 2$ are considered. The corresponding first-stage vectors χ, a, d and matrix G in (3.1)-(3.3) are as follows:

$$\begin{aligned} a^T &= \begin{pmatrix} F_1 & F_1 & F_1 & F_2 & F_2 & F_2 & q^1 & q^1 & q^1 & q^2 & q^2 & q^2 \end{pmatrix}, \\ \chi^T &= \begin{pmatrix} y_{11} & y_{21} & y_{31} & y_{12} & y_{22} & y_{32} & r_1^1 & r_2^1 & r_3^1 & r_1^2 & r_2^2 & r_3^2 \end{pmatrix}, \end{aligned}$$

$$G = \begin{pmatrix} 1 & & & 1 & & & & & & & & \\ & 1 & & & 1 & & & & & & & \\ & & 1 & & & 1 & & & & & & \\ -M_1 & & & -M_2 & & & b^1 & & & b^2 & & \\ & -M_1 & & & -M_2 & & & b^1 & & & b^2 & \\ & & -M_1 & & & -M_2 & & & b^1 & & & b^2 \end{pmatrix}, \quad d = \begin{pmatrix} 1 \\ 1 \\ 1 \\ 0 \\ 0 \\ 0 \end{pmatrix}$$

[17] See Appendix A.1 for an illustrative example of a piece-wise linear recourse function $Q(\chi, \xi(s))$. A proof of convexity can be found in Walkup and Wets [1967].

[18] The rest of this section is partially taken from Grass et al. [2018].

The first three rows in G belong to constraints (2.2) and the last three rows to (2.3).

For each scenario s the second-stage vectors in (3.4) take the following form:

$$\begin{aligned}
f^T &= \left(c^1_{(1,1)}\ c^2_{(1,1)}\ 0\ c^1_{(1,2)}\ c^2_{(1,2)}\ 0 \cdots c^2_{(3,3)}\ 0\ h^1\ h^1\ h^1\ h^2\ h^2\ h^2\ p^1\ p^1\ p^1\ p^2\ p^2\ p^2 \right),\\
\gamma_s^T &= \left(x^{1\,s}_{(1,1)}\ x^{2\,s}_{(1,1)}\ g_{(1,1)}\ x^{1\,s}_{(1,2)}\ x^{2\,s}_{(1,2)}\ g_{(1,2)} \cdots x^{2\,s}_{(3,3)}\ g_{(3,3)}\ z^1_1\ z^1_2\ z^1_3\ z^2_1\ z^2_2\ z^2_3\ w^1_1\ w^1_2\ w^1_3\ w^2_1\ w^2_2\ w^2_3 \right),\\
t_s^T &= \left(U^s_{(1,1)}\ U^s_{(1,2)}\ U^s_{(1,3)}\ U^s_{(2,1)}\ U^s_{(2,2)}\ U^s_{(2,3)}\ U^s_{(3,1)}\ U^s_{(3,2)}\ U^s_{(3,3)}\ v^{1s}_1\ v^{1s}_2\ v^{1s}_3\ v^{2s}_1\ v^{2s}_2\ v^{2s}_3 \right).
\end{aligned}$$

Since the arc capacity constraints in (2.4) are inequalities, slack variables $g_{(i,j)}$ for each arc $(i,j) \in R$ are inserted in the second-stage decision vector γ_s. The corresponding costs are represented by 0 in f. In the case of the Rawls and Turnquist [2010] model, the technology matrix T_s contains only ρ_i^{ks} which is connected to the first-stage pre-positioning variable r_i^k:

$$T_s = \begin{pmatrix} 0 & & & & \\ & \ddots & & & \\ & & 0 & & \\ & & & \rho_1^{1s} & & \\ & & & & \ddots & \\ & & & & & \rho_3^{2s} \end{pmatrix} \in \mathbb{R}^{(|N|\cdot|K|+|R|)\times|N|\cdot(|L|+|K|)}.$$

As will be shown later, the formulation of the recourse matrix W is decisive for constructing an efficient solution algorithm in terms of fast computation times. The specific form for the Rawls and Turnquist [2010] model formulation is:

$$W = \begin{pmatrix} u^T & & & & \\ & u^T & & & \\ & & \ddots & & \\ & & & u^T & \\ H_1 & H_2 & \cdots & H_{|R|} & Q \end{pmatrix} \in \mathbb{R}^{(|N|\cdot|K|+|R|)\times(|K|\cdot(|R|+2\cdot|N|)+|R|)}, \tag{3.5}$$

with $u^T = [u^1 \cdots u^{|K|}\ 1] \in \mathbb{R}^{1\times(|K|+1)}$ and

$$Q = \begin{bmatrix} -1 & & & 1 & & \\ & \ddots & & & \ddots & \\ & & -1 & & & 1 \end{bmatrix} \in \mathbb{R}^{|N|\cdot|K|\times 2\cdot|N|\cdot|K|}.$$

The recourse matrix W is fixed here, i.e. all parameters are independent of s and hence deterministic. In addition, it is a special *lower block triangular matrix*[19] with rectangular blocks, where all entries are zero except on the main block diagonal and the last block row. This structure is favorable for the solution method, as will be shown in Section 3.3.3. Note that the entry 1 in u corresponds to the slack variables $g_{(i,j)}$, as described above.

The last block row of W belongs to the flow conservation constraints (2.5) and contains block matrices $H_i \in \mathbb{R}^{|N|\cdot|K|\times(|K|+1)}$ for $i = 1, \ldots, |R|$ such that

[19]The name *block angular matrix* is also common in the literature.

$H = [H_1 \; H_2 \; \cdots H_{|R|}] \in \mathbb{R}^{|N|\cdot|K| \times |R|\cdot(|K|+1)}$. For the example given above, H is as follows:

$$H = \begin{pmatrix} 0&0&0&-1&0&0&-1&0&0&1&0&0&0&0&0&0&0&0&1&0&0&0&0&0&0&0&0 \\ 0&0&0&1&0&0&0&0&0&-1&0&0&0&0&0&-1&0&0&0&0&0&1&0&0&0&0&0 \\ 0&0&0&0&0&0&1&0&0&0&0&0&0&0&0&1&0&0&-1&0&0&-1&0&0&0&0&0 \\ 0&0&0&0&-1&0&0&-1&0&0&1&0&0&0&0&0&0&0&0&1&0&0&0&0&0&0&0 \\ 0&0&0&0&1&0&0&0&0&0&-1&0&0&0&0&0&-1&0&0&0&0&0&1&0&0&0&0 \\ 0&0&0&0&0&0&0&1&0&0&0&0&0&0&0&0&1&0&0&-1&0&0&-1&0&0&0&0 \end{pmatrix}.$$

For instance, the first row in H indicates if relief items of type 1 are supplied to node 1 (plus sign) or transported away from node 1 (minus sign), i.e. $-x^{1\,s}_{(1,2)} - x^{1\,s}_{(1,3)} + x^{1\,s}_{(2,1)} + x^{1\,s}_{(3,1)}$. Similarly, the second and third row represents flows of relief items of type 1 to and away from node 2 and 3, respectively. The last three rows in H are equivalent to the first three rows but apply for relief items of type 2.

How two-stage stochastic programs can be solved is discussed in the next section.

3.2.2 The L-Shaped Method

The L-shaped method [Slyke and Wets, 1969] is a classical approach for solving two-stage stochastic problems by exploiting the specific structure of the so-called *extensive form.*[20] Assuming a finite number of scenarios $s = 1, ..., |S|$, the two-stage stochastic program in (3.1) - (3.3) can be extended to its deterministic equivalent, the extensive form:

$$\min \quad a^T\chi \quad + \quad P_1 f^T\gamma_1 \quad + \quad P_2 f^T\gamma_2 \quad + \quad \dots \quad + \quad P_{|S|} f^T\gamma_{|S|} \tag{3.6}$$

$$s.t. \quad G\chi \qquad\qquad = d \tag{3.7}$$

$$\begin{array}{lllll} T_1\chi & + W\gamma_1 & & & = t_1 \\ T_2\chi & & + W\gamma_2 & & = t_2 \\ \vdots & & \ddots & & \vdots \\ T_{|S|}\chi & & & + W\gamma_{|S|} & = t_{|S|} \end{array} \tag{3.8}$$

$$\chi \geq 0 \tag{3.9}$$

$$\gamma_1, \gamma_2, ..., \gamma_{|S|} \geq 0. \tag{3.10}$$

This L-shaped structure can be used to decompose the entire problem (3.6)-(3.10) into a *master problem*:

$$\min \; a^T\chi + \sum_{s\in S} P_s f^T\gamma_s \quad \text{s.t. (3.7) and (3.9)}, \tag{3.11}$$

and $|S|$ *sub-problems* of the second stage, which are defined in (3.18) below. The (non-linear) expected recourse function $E_\xi\left[Q(\chi, \xi(s))\right]$ in (3.1) is replaced by $\sum_{s\in S} P_s f^T\gamma_s$. The

[20]The following text (to page 52) is partially taken from Grass and Fischer [2016b].

latter can now be approximated by θ such that the general two-stage stochastic program (3.1)-(3.3) can be restated as:

$$\min\ a^T\chi + \theta \tag{3.12}$$
$$\text{s.t. } G\chi = d \tag{3.13}$$
$$\theta \geq E_\xi\left[Q(\chi, \xi(s))\right] \tag{3.14}$$
$$\chi \geq 0. \tag{3.15}$$

The main idea of the L-shaped method is to substitute (3.14) by successively adding so-called *optimality cuts* [Birge and Louveaux, 2011, p. 190]. These cuts represent additional constraints restricting the solution space. The general procedure of the L-shaped method including the optimality cuts is described in Figure 3.2.[21]

In the initialization step, i.e. $\iota = 0$ in Figure 3.2, the master problem (3.11) is solved, initially neglecting $\sum_{s\in S} P_s f^T \gamma_s$. The corresponding first-stage solution $\overline{\chi}^0$ is then treated as an input parameter for the sub-problems of the second stage. In particular, the primal sub-problem for each scenario s is given in (3.18). Using slack variables z_s the corresponding dual problem can be written as

$$\max(t_s - T_s\overline{\chi}^\iota)^T\pi_s \quad \text{s.t.} \quad W^T\pi_s + z_s = f \quad z_s \geq 0, \tag{3.16}$$

where π_s is the vector of dual variables and is also referred to as the vector of simplex multipliers.[22]

[21] Figure 3.2 and the following text (to page 52) are partially taken from Grass et al. [2018].

[22] In the classical version of the L-shaped method, as shown in Figure 3.2, it is sufficient to solve the primal sub-problem (3.18). Based on, e.g., the final tableau of the simplex method the shadow prices can be obtained representing the required simplex multipliers $\overline{\pi}_s^\iota$ [Diwekar, 2008, p. 23]. However, the dual problem in (3.16) will become relevant in Section 3.3.3 where a primal-dual interior-point method is introduced providing $\overline{\gamma}_s^\iota$ and $\overline{\pi}_s^\iota$ simultaneously.

Step 0: Initialization

- Set iteration counter $\iota = 0$ and solve master problem (3.11) without $\sum_{s \in S} P_s f^T \gamma_s$
- Result: $\overline{\chi}^0$
- Set $\theta^0 = -\infty$ and go to Step 2

Step 1: Solving Master Problem

- Solve master problem

$$\begin{aligned} &\min a^T \chi + \theta \\ &\text{s.t.} \quad G\chi = d \\ &\theta \geq \sum_{s \in S} P_s (t_s - T_s \chi)^T \overline{\pi}_s^{\kappa} \quad \forall \kappa = 0, ..., \iota - 1 \\ &\chi \geq 0, \theta \in \mathbb{R} \end{aligned} \tag{3.17}$$

- Result: $\overline{\chi}^{\iota}$ and θ^{ι}
- Go to Step 2

Step 2: Solving Sub-Problems

- For $s = 1, ..., |S|$ solve primal sub-problem

$$\min f^T \gamma_s \quad \text{s.t.} \quad W\gamma_s = t_s - T_s \overline{\chi}^{\iota} \quad \gamma_s \geq 0 \tag{3.18}$$

- Result: $\overline{\gamma}_s^{\iota}$ and the vector of simplex multipliers $\overline{\pi}_s^{\iota}$
- Calculate $\vartheta^{\iota} = \sum_{s \in S} P_s (t_s - T_s \overline{\chi}^{\iota})^T \overline{\pi}_s^{\iota}$

Step 3: Termination

- If $\vartheta^{\iota} \leq \theta^{\iota}$ STOP, optimal solution found
  Otherwise set $\iota = \iota + 1$, add optimality cut (3.17) to the master problem and go to Step 1.

Figure 3.2: The L-shaped method (based on Birge and Louveaux [2011, Ch. 5])

After solving the sub-problems (3.18) the objective function value of the second-stage sub-problem $\vartheta^0 = \sum_{s \in S} P_s (t_s - T_s \overline{\chi}^0)^T \overline{\pi}_s^0$ is determined, using the vector of simplex multipliers $\overline{\pi}_s^0$ (see Step 2 in Figure 3.2). If ϑ^0 exceeds the current θ^0, the iteration counter

ι is increased by one and a first optimality cut (3.17)

$$\theta \geq \sum_{s \in S} P_s (t_s - T_s \chi)^T \overline{\pi}_s^0$$

is inserted to the master problem. For $\iota = 1$ the master problem in Step 1 of the algorithm is solved, resulting in an updated first-stage solution $\overline{\chi}^\iota$. Afterwards, $\overline{\chi}^\iota$ is again used for solving the sub-problems, leading to an additional optimality cut (3.17). An example of finitely many optimality cuts is given in Figure 3.3. Here, the minimum of the expected recourse function $E_\xi \left[Q(\chi, \xi(s))\right]$ is obtained after four cuts. The aim is to approximate the expected recourse function from below by means of optimality cuts.

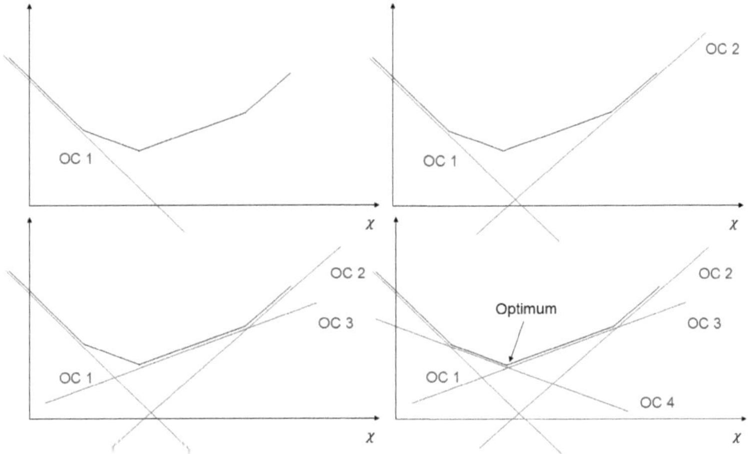

Figure 3.3: Optimality cuts (OC) approximating the expected recourse function $E_\xi \left[Q(\chi, \xi(s))\right]$

Note that χ in (3.17) is again a vector of decision variables whereas $\overline{\pi}_s^\kappa$ are the simplex multipliers found in previous iterations $\kappa = 0, ..., \iota - 1$. As long as the solution is not optimal, i.e. $\theta^\iota < \vartheta^\iota$ for the current $\overline{\chi}^\iota$, the process of solving the master problem and sub-problems successively has to be repeated.

In the case of Rawls and Turnquist [2010], the master problem is:

$$\min \sum_{i \in \mathrm{I}} \sum_{l \in \mathrm{L}} \mathrm{F}_l y_{il} + \sum_{k \in \mathrm{K}} \sum_{i \in \mathrm{I}} \mathrm{q}^k r_i^k + \theta, \tag{3.19}$$

s.t. (2.2), (2.3), (2.6) and (2.7). For each scenario $s = 1, ..., |S|$ the corresponding sub-problem has the form:

$$\min \sum_{(i,j) \in \mathrm{R}} \sum_{k \in \mathrm{K}} \mathrm{c}_{(i,j)}^k x_{(i,j)}^{ks} + \sum_{i \in \mathrm{I}} \sum_{k \in \mathrm{K}} (\mathrm{h}^k z_i^{ks} + \mathrm{p}^k w_i^{ks}),$$

s.t. (2.4), (2.5) with given first-stage variables r_i^k from the master problem and the non-negativity constraints (2.8) and (2.9).

The following theorem states that the L-shaped method provides an exact solution after a finite number of iterations [Birge and Louveaux, 2011, p. 196].

Theorem 3.2.1. *The L-shaped method has the finite convergence property for problems of the form (3.6)-(3.10).*

Proof. [23] From duality theory it is known that the primal and dual objective function values in (3.18) and (3.16), respectively, are equal at the optimum:

$$f^T \overline{\gamma}_s = Q(\overline{\chi}^\iota, \xi(s)) = (t_s - T_s \overline{\chi}^\iota)^T \overline{\pi}_s^\iota \quad \forall s \in S. \tag{3.20}$$

This holds for a given first-stage solution $\overline{\chi}^\iota$ and the corresponding optimal primal and dual sub-problem solutions $\overline{\gamma}_s$ and $\overline{\pi}_s^\iota$, respectively. Since the recourse function $Q(\chi, \xi(s))$ in (3.4) is convex but non-differentiable at $\overline{\chi}$, the sub-gradient instead of the gradient has to be used. As shown in Figure 3.4 more than one sub-gradient exists at the non-differentiable point $\hat{x}_2$.

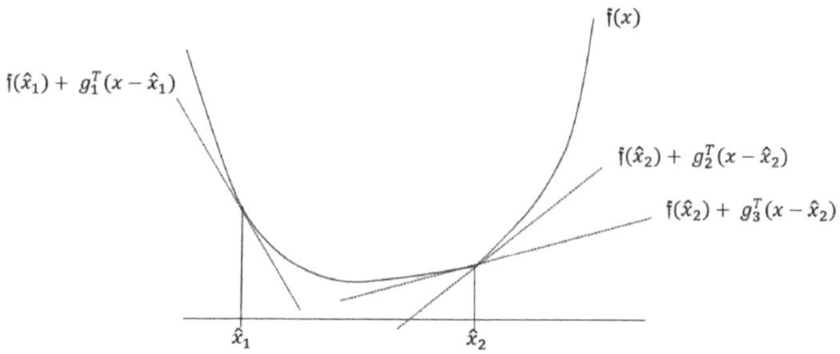

Figure 3.4: Sub-gradients of a non-smooth function (based on Boyd and Vandenberghe [2004, p. 69])

In general, a vector $g \in \mathbb{R}^n$ is a sub-gradient of a function $\mathfrak{f} : \mathbb{R}^n \longrightarrow \mathbb{R}$ at point $\hat{x} \in \mathsf{dom}\,\mathfrak{f}$ if [24]

$$\mathfrak{f}(x) \geq \mathfrak{f}(\hat{x}) + g^T(x - \hat{x}) \quad \forall x \in \mathsf{dom}\,\mathfrak{f}, \tag{3.21}$$

where

$$\mathsf{dom}\,\mathfrak{f} = \{x \in \mathbb{R}^n : \mid \mathfrak{f}(x) \mid < \infty\}$$

[23] The following proof is based on Birge and Louveaux [2011, pp. 196].

[24] See for example Nesterov (2013).

is the domain of the function $\mathfrak{f}$, i.e. the function value for an arbitrary x is finite. The sub-differential $\partial\mathfrak{f}(\hat{x})$ is the set of all sub-gradients of $\mathfrak{f}$ at $\hat{x}$. Here, the sub-gradient at $\overline{\chi}^\iota$ of (3.20) is of the form:

$$g(\overline{\chi}^\iota) = -T_s^T \overline{\pi}_s^\iota \in \partial Q(\overline{\chi}^\iota, \xi(s)), \quad \forall s \in S.$$

It follows from (3.21):

$$Q(\chi, \xi(s)) \geq Q(\overline{\chi}^\iota, \xi(s)) - (T_s^T \overline{\pi}_s^\iota)^T (\chi - \overline{\chi}^\iota) \quad \forall s \in S.$$

Using (3.20), $Q(\overline{\chi}^\iota, \xi(s))$ can be replaced by $(t_s - T_s\overline{\chi}^\iota)^T \overline{\pi}_s^\iota$

$$Q(\chi, \xi(s)) \geq (t_s - T_s\overline{\chi}^\iota)^T \overline{\pi}_s^\iota - (T_s^T \overline{\pi}_s^\iota)^T (\chi - \overline{\chi}^\iota) \quad \forall s \in S. \tag{3.22}$$

As two terms cancel out, (3.22) reduces to

$$Q(\chi, \xi(s)) \geq (t_s - T_s\chi)^T \overline{\pi}_s^\iota. \tag{3.23}$$

Finally, taking the expectation of (3.23) leads to

$$E_\xi \left[Q(\chi, \xi(s))\right] \geq E_\xi \left[(t_s - T_s\chi)^T \overline{\pi}_s^\iota\right] = \sum_{s \in S} P_s (t_s - T_s\chi)^T \overline{\pi}_s^\iota. \tag{3.24}$$

Due to (3.14) and (3.24), the following inequality holds

$$\theta \geq \sum_{s \in S} P_s (t_s - T_s\chi)^T \overline{\pi}_s^\iota,$$

which is exactly the optimality cut defined in (3.17). At the optimum $\theta^\iota = E_\xi\left[Q(\overline{\chi}^\iota, \xi(s))\right]$, which implies that $\theta^\iota = \sum_{s \in S} P_s (t_s - T_s\overline{\chi}^\iota)^T \overline{\pi}_s^\iota$, based on (3.20). Therefore, the L-shaped method terminates if $\theta^\iota = \vartheta^\iota$. As long as the current costs ϑ^ι for a given first-stage solution $\overline{\chi}^\iota$ exceed θ^ι, i.e. $\theta^\iota < \vartheta^\iota$, additional optimality cuts of the form (3.17) have to be added. In this case, $|S|$ new multipliers $\overline{\pi}_s^\iota$ are generated, which differ from the ones determined by earlier iterations.[25] The problem in (3.18) has finitely many feasible bases corresponding to the recourse matrix W. Since the vector $\overline{\pi}_s^\iota$ belongs to one of these bases, only finitely many optimality cuts (3.17) exist. As a result, the L-shaped method converges after a finite number of iterations. □

The Multi-Cut Approach

In the classical version of the L-shaped method, a single optimality cut (3.17) is generated by summing up the multipliers $\overline{\pi}_s^\kappa$ over all scenarios, potentially losing valuable information.[26]

[25]Due to the successively added optimality cuts (3.17) the first-stage solution $\overline{\chi}^\iota$ is updated in every iteration leading to a constantly updated $\overline{\pi}_s^\iota$.

[26]The following text is partially taken from Grass and Fischer [2016b] and Grass et al. [2018].

In the multi-cut approach one cut per scenario s is added to the master problem such that θ in (3.12) is replaced by θ_s:

$$\min a^T\chi + \sum_{s\in S} \theta_s.$$

Hence, optimality cuts (3.17) are changed to

$$\theta_s \geq P_s(t_s - T_s\chi)^T \overline{\pi}_s^{\kappa} \quad \forall \kappa = 0, ..., \iota - 1; \forall s = 1, ..., |S|. \tag{3.25}$$

In this way, several optimality cuts are added to the master problem in every iteration, decreasing the required number of iterations, but also leading to an increased size of the master problem. As stated by Birge and Louveaux [2011, p. 200], the multi-cut approach is recommended if the number of scenarios does not significantly exceed the number of first-stage constraints.

For instance, Noyan [2012] proposes two algorithms for the generation of optimality cuts for the cVaR term in (2.10). In both algorithms single optimality cuts are replaced by multi-cuts (3.25). In order to overcome the increased problem size due to increased number of cuts, Noyan et al. [2016] utilize the so-called *lazy constraint callback* of CPLEX [Rubin, 2011]: In the standard L-shaped method described above, the master problem has to be solved every time an optimality cut is added. Therefore, it may happen that the same solutions are reevaluated again. To avoid such repeated evaluations the callback approach constructs a single branch-and-cut search tree where every node is visited only once. During this solution process the optimality and feasibility cuts generated by the L-shaped method are saved in a pool and renamed "lazy constraints". Only if CPLEX detects a potential violation of certain lazy constraints these constraints are added to the master problem. This approach leads to a reduction of computing effort and storage requirement. For a further reduction of computation time, Noyan et al. [2016] solve the sub-problems in parallel. Another idea for the acceleration of the standard L-shaped method is proposed in the following section.

3.3 The Accelerated L-Shaped Method

3.3.1 The Basic Idea

Although the algorithms reviewed in Section 3.1.1 exploit the specific structure in (3.6)-(3.10) and make use of enhanced techniques or adjusted heuristics, the computational effort becomes prohibitive if large-scale problems have to be solved to optimality.[27] For these cases, heuristics seem to be the only way to find solutions within a reasonable time frame. Despite their widespread use heuristics have some serious drawbacks. To begin with, heuristically found solutions are usually not exact. The added value of exact solutions cannot really be determined due to the potentially imprecise estimations of input data in

[27]The next paragraph is partially taken from Grass and Fischer [2016b].

the uncertain environment of disasters. However, solution quality can be decisive since inaccurate input parameters in combination with inaccurate solutions may lead to wrong decisions, resulting in the loss of human lives. In addition to the inexactness of solutions, heuristics are characterized by the general inability to measure the solution quality in terms of deviations from optimality. Another disadvantage of heuristics is the frequent necessity of parameter adjustments. For instance, the crossover rate in evolution-based heuristics (see Section 3.1.2) has to be adjusted to the input data and to the problem at hand. Moreover, most heuristics suffer from their limited flexibility in adapting to different model structures. A heuristic which performs well for a specific problem may provide solutions of lower quality when applied to another problem. In this case appropriate modifications are needed. Modifying heuristics and manipulating parameters require a special kind of expertise. It is questionable if aid organizations have the appropriate knowledge or have the financial and temporal resources to acquire it. Hence, the development of a solution method which meets the requirements of relief organizations is essential. In particular, the solution method should be applicable to different problems without parameter modifications and provide robust solutions regardless of underlying input data. Moreover, a solution of adequate quality available within reasonable computing time is essential when emergency situations occur.

Against this background, an accelerated L-shaped method is developed which is applicable to a whole class of two-stage stochastic programs and provides solutions much faster than a commercial solver, especially for realistic large-scale problems.[28] The main challenge when developing an accelerated version of the L-shaped method is the efficient solution of the master problem in Step 1 and the sub-problems in Step 2 of the algorithm in Figure 3.2. Often integer or binary decision variables occur only on the first stage in the relevant literature, e.g. decisions concerning locations as in Rawls and Turnquist [2010]. In contrast, second-stage sub-problems are in general "simple" LP problems. However, these sub-problems can be very large in comparison to the master problem. For instance, assuming the entire American southeast coast with its 900 counties has to be prepared for the upcoming hurricane season, i.e. $|N| = |I| = 900$, such that the network contains potentially 810,000 shortest paths between each pair of nodes. While the actual network will likely be sparser, $|R| = 810,000$ arcs may exist in the worst case. If the number of item types and facility sizes is set to $|K| = |L| = 3$, the master problem for the Rawls and Turnquist [2010] model contains $|I| \cdot (|K| + |L|)$ =5,400 decision variables (of which 2,700 are binary) and $2 \cdot |I|$ =1,800 constraints. In contrast, each sub-problem has $|R| \cdot |K| + 2 \cdot |N| \cdot |K| + |R|$ =3,245,400 variables and $|R| + |N| \cdot |K|$ =812,700 constraints. Since $|S|$ sub-problems have to be solved in each iteration of the L-shaped method, solving these problems efficiently is crucial for the overall performance.[29] Usually the linear sub-problems are solved via the simplex algorithm [Rahmaniani et al., 2017]. However, as the problem size increases, the computational effort of such exact methods

[28]The rest of this section is partially taken from Grass et al. [2018].

[29]Note that the sub-problems can also be solved in parallel, accelerating the L-shaped procedure further.

becomes prohibitive.[30] As an alternative, iterative algorithms could be applied providing approximate solutions. In particular, interior-point methods are especially popular for large-scale problems [Gondzio, 2012].

Based on this discussion, the accelerated L-shaped method for solving two-stage stochastic programs is as follows:

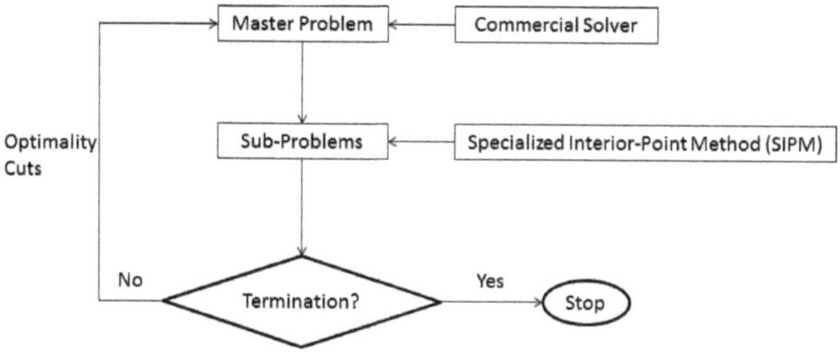

Figure 3.5: The accelerated L-shaped method [Grass et al., 2018]

As shown in Figure 3.5, the master problem is solved via a commercial solver like CPLEX or Gurobi.[31] Since master problems in the humanitarian context are often of moderate size, e.g. 5,400 variables as in the example above, they do not represent a significant hurdle for commercial optimization software. In order to solve large-scale sub-problems efficiently a *specialized interior-point method* (SIPM) is proposed which exploits the specific structure of the recourse matrix W in (3.5). The proposed solution procedure in Figure 3.5 is the same as the standard L-shaped method but where the sub-problems are solved via the specialized interior-point method.

Although interior-point methods are often used for two-stage stochastic programs [Rahmaniani et al., 2017], to the knowledge of the author, the specific structure inherent in the recourse matrix W has not been exploited so far. For instance, Birge and Holmes (1992) directly make use of the block structure of the extensive form (3.6)-(3.10) to develop an efficient interior-point algorithm. In contrast to the L-shaped method, a decomposition into a master problem and sub-problems is not performed. Their approach is applicable for two-stage stochastic problems containing solely continuous decision variables. Since most of the real-world problems are modeled as (mixed) integer programs the applicability of their approach is limited.

Before describing the specialized interior-point method in more detail (Section 3.3.3), necessary assumptions for the use of the accelerated L-shaped algorithm are stated first.

[30] In the worst-case the simplex algorithm requires 2^n iterations to solve linear programs, where n is the number of decision variables [Vanderbei, 2014, p. 42].

[31] More information on CPLEX and Gurobi can be found on https://www.ibm.com/us-en/marketplace/ibm-ilog-cplex and http://www.gurobi.com/, respectively.

3.3.2 Assumptions

The solution method proposed in Figure 3.5 is applicable to a whole class of two-stage stochastic programs in disaster management, fulfilling the following assumptions:

1. Size of the master problem is relatively small
2. Recourse matrix W has a block triangular structure as in (3.5)
3. Standard L-shaped method is applicable

The first requirement is often met for two-stage stochastic problems in humanitarian context. For instance, decisions on locations are the most common first-stage decisions (see Table 2.2). As the number of potential disaster locations is generally modest, the corresponding location problem and therefore the master problem is computationally manageable. The same holds for first-stage decisions concerning, e.g., the supplier selection [Bai, 2016], or bridge retrofitting [Liu et al., 2009]. A commercial solver is a reasonable choice for tackling such master problems as shown in Figure 3.5.

According to the second assumption, the recourse matrix W should have a block triangular structure as in (3.5). Since it is tedious to set up W by hand, an automated procedure would be helpful, making the accelerated L-shaped method more user-friendly. Fortunately, several ordering techniques exist in the literature to construct block triangular matrices similar to (3.5). For instance, Hu et al. [2000] propose, in the context of chemical process simulation, a scheme where any[32] sparse matrix can be reordered such that the required structure is obtained. In the case of two-stage stochastic programs, the linking constraints, like the flow conservation constraints in (2.5), can form the last block row of W. Pure second-stage constraints with solely second-stage variables, e.g. the arc capacity constraints in (2.4), can be used for the diagonal blocks. Such a reordering algorithm can be part of the preprocessing stage where input data are automatically prepared for the main solution procedure. In this case, relief organizations do not have to form a block triangular W by themselves.

The third assumption seems rather restrictive as the standard L-shaped method cannot be used if integer second-stage variables occur in the sub-problems. For this purpose interior-point methods have to be adjusted and cannot be applied as proposed here. However, most of the second-stage problems in the humanitarian literature are concerned with transportation issues (see Table 2.2 and 2.4) and represent linear programs. Table 3.3 summarizes the model types of the second-stage problems reviewed in Section 2.2.1. If the second-stage problem contains only continuous variables, the corresponding model type is classified as a linear program (LP), otherwise as a mixed integer linear program (MILP).

[32] For instance, symmetry as well as specific properties or structures of the matrix are not required.

Table 3.3: Model type of the second stage

Author(s)	Year	Second Stage LP	MILP
Barbarosoglu and Arda	2004	x	
Chang et al.	2007	x	
Balcik and Beamon	2008	x	
Andreas and Smith	2009	x	
Liu and Huang	2009	x	
Liu et al.	2009	x	
Ng and Waller	2009	x	
Fan and Liu	2010	x	
Mete and Zabinsky	2010	x	
Peeta et al.	2010		x
Rawls and Turnquist	2010	x	
Salmeron and Apte	2010		x
Cai et al.	2011	x	
Falasca and Zobel	2011	x	
Li et al.	2011	x	
Rawls and Turnquist	2011		x
Bozorgi-Amiri et al.	2012	x	
Döyen et al.	2012	x	
Li et al.	2012		x
Lodree et al.	2012	x	
Noyan	2012	x	
Rawls and Turnquist	2012		x
Tricoire et al.	2012	x	
Wang et al.	2012	x	
Bozorgi-Amiri et al.	2013	x	
Davis et al.	2013		x
Klibi et al.	2013	x	
Maqsood and Huang	2013	x	
Wang and Huang	2013	x	
Ahmadi et al.	2015	x	
Hong et al.	2015	x	
Hu et al.	2015	x	
Verma and Gaukler	2015	x	
Zolfaghari and Peyghaleh	2015	x	
Alem et al.	2016		x
Bai	2016	x	
Noyan et al.	2016		x
Rath et al.	2016	x	
Rezaei-Malek et al.	2016	x	
Tofighi et al.	2016	x	

As indicated in Table 3.3 most, namely 80%, of the sub-problems considered here are linear programs. Interior-point methods are therefore an appropriate choice for solving large-scale sub-problems, which are often proposed in the humanitarian literature.

It should be mentioned that for problems where the recourse is not complete, it may be necessary to insert feasibility cuts in addition to the optimality cuts. In this way, the feasibility of all sub-problems for arbitrary first-stage solutions χ is guaranteed [Birge and Louveaux, 2011, pp. 191]. For instance, feasibility cuts are needed in the model of Liu et al. [2009], where the sub-problem is formulated as a multi-commodity network flow problem. For problems with incomplete recourse the solution approach proposed in Figure 3.5 is still applicable with the exception that the master problem contains two types of cuts. Hence, solving the sub-problems via the specialized interior-point method SIPM is not affected by such feasibility considerations. Often two-stage stochastic programs can even be modified to achieve complete recourse.[33]

In the following SIPM is proposed for solving large-scale sub-problems in disaster management.

3.3.3 Specialized Primal-Dual Interior-Point Method

Especially for large-scale problems, interior-point methods often outperform the simplex algorithm [Ferris et al., 2007, p. 212].[34] Instead of evaluating vertices of the feasible region until the optimal solution is found, interior-point methods reach the optimal vertex from the interior. As a result, the algorithm can be terminated when the solution is close enough to the boundary, resulting in an approximate solution. This is particularly beneficial for the L-shaped method, since exactness is not required for the construction of valid optimality cuts (3.17) [Rahmaniani et al., 2017].

Before showing how the specific structure of W in (3.5) can be exploited by the specialized interior-point method SIPM for solving sub-problems (3.18) and (3.16), a standard interior-point algorithm, namely a *primal-dual path-following method*, is described below.[35]

A Primal-Dual Path-Following Method

The general procedure is to remove the non-negativity constraint $\gamma \geq 0$ from the primal sub-problem (3.18) by formulating the *logarithmic barrier function*:

$$\min\ f^T\gamma - \mu\sum_{j=1}^{n_2}\log\gamma_j \tag{3.26}$$

$$s.t.\ W\gamma = \tilde{t}, \tag{3.27}$$

[33]See Birge and Louveaux [2011, p. 193] for corresponding techniques.

[34]The following section is partially taken from Grass et al. [2018].

[35]Unless otherwise stated, W represents an arbitrary matrix in the following, i.e. does not necessarily possess the structure in (3.5). Since the approach is presented for one sub-problem, the subscript s can be omitted for now.

where $\mu > 0$ is a barrier term, $\tilde{t} = t - T\overline{\chi}^{\iota}$ and $n_2 = |R| \cdot |K| + 2 \cdot |N| \cdot |K| + |R|$ for the reference model of Rawls and Turnquist [2010] in (2.1)-(2.9). Note that the logarithm is not defined for non-positive real numbers and goes to negative infinity as γ_j goes to zero, i.e. increasing the objective value (3.26). For μ close to zero, the original problem is approximated.

Adding the equality constraint (3.27) to the objective function (3.26) leads to the corresponding *Lagrangian function*

$$L(\gamma, \pi) = f^T\gamma - \mu \sum_{j=1}^{n_2} \log \gamma_j + \pi^T(\tilde{t} - W\gamma), \tag{3.28}$$

where $\pi \in \mathbb{R}^{m_2}$ is the vector of Lagrange multipliers and at the same time the vector of dual variables in (3.16). At the optimum the Karush–Kuhn–Tucker (KKT) first-order optimality conditions should be met [Wright, 1997, p. 240]:

$$\nabla_\gamma L = f - \mu\Gamma^{-1}e - W^T\pi = 0 \tag{3.29}$$
$$\nabla_\pi L = \tilde{t} - W\gamma = 0 \tag{3.30}$$

with $e = (1, ..., 1)^T$, $\Gamma = diag(\gamma_1, ..., \gamma_{n_2})$ and 0 as a vector of appropriate size. By setting $z = \mu\Gamma^{-1}e$ the first-order optimality conditions (3.29)-(3.30) can be reformulated as:

$$W^T\pi + z = f \tag{3.31}$$
$$W\gamma = \tilde{t} \tag{3.32}$$
$$\Gamma Z e = \mu e \tag{3.33}$$
$$\gamma, z > 0 \tag{3.34}$$

with $Z = diag(z_1, ..., z_{n_2})$. Equation (3.31) states the dual and (3.32) the primal feasibility. Note that for $\mu = 0$ the standard complementarity conditions are obtained in (3.33), i.e. at the optimum either γ_j or z_j has to be zero (for $j = 1, ..., n_2$). In this case the *primal-dual solution* vector is (γ^*, π^*, z^*) if γ^* is the solution of the primal problem (3.18) and (π^*, z^*) the solution of the dual problem (3.16).

The main idea of interior-point methods is to perturb the standard complementarity conditions $\Gamma Z e = 0$ by introducing the barrier parameter $\mu > 0$ in (3.33). During the solution process μ is successively reduced. In this way the optimal solution is approached from the interior of the feasible set following the so-called *primal-dual central path* $\mathcal{C}$:

$$\mathcal{C} = \{(\gamma_\mu, \pi_\mu, z_\mu) \mid \mu > 0\},$$

where $(\gamma_\mu, \pi_\mu, z_\mu)$ represent the solutions of (3.31)-(3.34) for a specific choice of barrier parameter μ. In Figure 3.6 a feasible set for primal variables γ is shown in the form of a polygon with an exemplary central path $\mathcal{C}$.

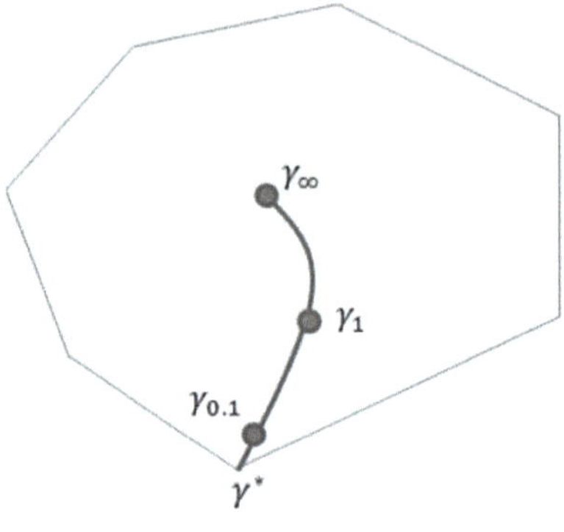

Figure 3.6: The central path $\mathcal{C}$ with γ_μ, depending on different values of μ, i.e. $\mu = \infty$, $\mu = 1$ and $\mu = 0.1$ (based on Kwon [2013, p. 223])

As one can see in Figure 3.6, the interior point starts theoretically from the *analytic center* ($\mu = \infty$) and approaches the optimal solution (bottom vertex on the boundary) of the original problem (3.18) and (3.16) as $\mu \to 0$. In other words, for each μ the solutions to (3.31)-(3.34) define one point on the central path $\mathcal{C}$, which lies strictly inside the feasible set, i.e. where γ and z are strictly positive.

Due to the nonlinearity of (3.33), solving system (3.31)-(3.34) exactly is computationally expensive. Therefore, interior-point algorithms generally apply the Newton method[36] to generate approximate solutions to (3.31)-(3.34). In doing so, points are generated in each iteration which follow the central path $\mathcal{C}$ only approximately. This is the reason for the name *path-following methods*.

Using Newton's method for the system (3.31)-(3.33), it can be solved iteratively with search directions determined through:

$$\begin{aligned} W^T \triangle\pi + \triangle z &= f - W^T\pi - z \\ W\triangle\gamma &= \tilde{t} - W\gamma \\ Z\triangle\gamma + \Gamma\triangle z &= -\Gamma Z e + \mu e, \end{aligned}$$

with $\triangle$ indicating the Newton search direction. From now on, γ, π, z are given, whereas $\triangle\gamma, \triangle\pi$ and $\triangle z$ have to be determined representing the respective search direction. The corresponding matrix notation is:

$$\begin{bmatrix} 0 & W^T & I \\ W & 0 & 0 \\ Z & 0 & \Gamma \end{bmatrix} \begin{bmatrix} \triangle\gamma \\ \triangle\pi \\ \triangle z \end{bmatrix} = \begin{bmatrix} f - W^T\pi - z \\ \tilde{t} - W\gamma \\ -\Gamma Z e + \mu e \end{bmatrix}, \tag{3.35}$$

[36]See Appendix A.2 for a general description of Newton's method for solving systems of non-linear equations.

where I is an identity matrix of appropriate size. The right hand side in (3.35) represents the residuals to the corresponding constraints (3.31), (3.32) and (3.33), respectively.

A crucial step of interior-point methods is the choice of an appropriate barrier parameter μ which influences the search for an optimal solution. The parameter μ should be chosen in such a way that a long step in the Newton direction can be made, but such that the positivity conditions $\gamma > 0$ and $z > 0$ are still fulfilled. This can be achieved if the iterates lie in the center of the feasible set, i.e. near the central path $\mathcal{C}$. Therefore, the barrier parameter is generally set to $\mu = \sigma\tau$, where $\sigma \in (0,1)$ represents the *centering parameter* and

$$\tau = \frac{1}{n_2}\sum_{j=1}^{n_2} \gamma_j z_j = \frac{\gamma^T z}{n_2}, \tag{3.36}$$

is called the *duality measure*[37] [Nocedal and Wright, 2006, p. 395]. The name originates from the fact that the product $n_2\tau$ measures the duality gap, i.e. the difference between the primal objective value $f^T\gamma$ in (3.18) and dual objective value $(t - T\overline{\chi}^{\iota})^T\pi$ in (3.16):

$$n_2\tau = \gamma^T z = \gamma^T(f - W^T\pi) = f^T\gamma - (W\gamma)^T\pi = f^T\gamma - (t - T\overline{\chi}^{\iota})^T\pi.$$

In each iteration of path-following interior-point methods the duality gap is reduced successively until the optimal solution is found. The reduction process is influenced by the centering parameter σ which has to be chosen carefully. In the case where the centering parameter σ is close to 1, μ corresponds approximately to the mean value τ in (3.36). According to the last row in (3.35), the Newton direction is computed such that the next iterate is forced to be closer to the central path $(\gamma_\tau, \pi_\tau, z_\tau) \in \mathcal{C}$ where each product $\gamma_j z_j$, $j = 1, ..., n_2$ in (3.33) is nearly equivalent to the *current* mean value τ, i.e. $\gamma_j z_j = \gamma^T z / n_2$ for $j = 1, ..., n_2$. As a result, τ in (3.36) remains almost unchanged in the next iteration. As $\mu \approx \tau$, for $\sigma \to 1$, neither the optimality gap, measured by τ, nor the complementarity gap, measured by μ, can be reduced significantly. However, the advantage of the new iterate is that it is located nearer to the central path. A well centered iterate can take a longer Newton step toward the optimum in the next iteration without leaving the feasible set.

In contrast, for $\sigma \to 0$ the barrier parameter μ is close to 0 such that the complementarity conditions in (3.33) are barely disturbed, leading to an iterate which is closer to the optimal solution than the previous iterate. As each product $\gamma_j z_j$, $j = 1, ..., n_2$ in (3.33) is approximately 0, the duality measure τ in (3.36) is also near 0. However, choosing a small centering parameter σ too early, i.e. when the iterate is still far away from optimum, only a short step in the Newton direction may be taken in the next iteration.

As a result, σ should be chosen such that the new point is closer to the central path $\mathcal{C}$, i.e. $\sigma \to 1$, and at the same time has a smaller duality measure τ, i.e. $\sigma \to 0$. According to Vanderbei [2014, p. 272], $\sigma = 0.1$ has been proven appropriate in practice. In contrast, Ferris et al. [2007, p. 210] choose σ depending on the duality measure τ,

[37]The term *mean value* is also often used in the relevant literature.

namely $\sigma = \min\{0.1,\ 100\,\tau\}$. If the duality gap is large, e.g. at the beginning of the search process, σ is set to 0.1 leading to a better centering and therefore allowing a long step toward the Newton direction in the next iteration. On the other hand, for a small duality measure τ, i.e. when the optimal solution is almost approached, a point near the central path $\mathcal{C}$ is not needed anymore.

In Figure 3.7, the pseudo code of a primal-dual path-following interior-point method is given.

Step 0: Initialization

- Choose (γ^0, π^0, z^0) such that $\gamma^0 > 0$ and $z^0 > 0$
- Set $\mathbf{k} = 1$ (iteration counter)

Step 1: Determination of Search Direction

- Set $\tau^{\mathbf{k}} = \frac{(\gamma^{\mathbf{k}-1})^T z^{\mathbf{k}-1}}{n_2}$ (duality measure)
- Set $\sigma^{\mathbf{k}} = \min\{0.1,\ 100\,\tau^{\mathbf{k}}\}$ (centering parameter)
- Set $\mu^{\mathbf{k}} = \sigma^{\mathbf{k}}\tau^{\mathbf{k}}$ (barrier parameter)
- Solve (3.35) to determine search direction $(\triangle\gamma^{\mathbf{k}}, \triangle\pi^{\mathbf{k}}, \triangle z^{\mathbf{k}})$

Step 2: Update Solution

- Set $\eta^{\mathbf{k}} = \max\{0.9995, 1 - \tau^{\mathbf{k}}\}$ (step length dampening parameter)
- Compute step length $\alpha^{\mathbf{k}}_{\gamma,max}$ and $\alpha^{\mathbf{k}}_{z,max}$ according to (3.37)
- Choose step lengths $\alpha^{\mathbf{k}}_{\gamma} = \min\{1, \eta^{\mathbf{k}}\,\alpha^{\mathbf{k}}_{\gamma,max}\}$ and $\alpha^{\mathbf{k}}_{z} = \min\{1, \eta^{\mathbf{k}}\,\alpha^{\mathbf{k}}_{z,max}\}$ such that the new iterate $\gamma^{\mathbf{k}}, z^{\mathbf{k}} > 0$
- Update solution $(\gamma^{\mathbf{k}}, \pi^{\mathbf{k}}, z^{\mathbf{k}})$ according to (3.38)

Step 3: Termination

- If convergence criteria (3.39) is fulfilled STOP, otherwise set $\mathbf{k} = \mathbf{k} + 1$ and go to Step 1

End

Figure 3.7: A primal-dual path-following interior-point method

After choosing an interior starting point (γ^0, π^0, z^0), i.e. γ^0 and z^0 are strictly positive (Step 0), the main process is executed (Step 1 to Step 3). First of all the duality measure $\tau^{\mathbf{k}}$ and centering parameter $\sigma^{\mathbf{k}}$ are determined (Step 1). Based on these values a barrier parameter $\mu^{\mathbf{k}}$ and afterwards the search direction $(\triangle\gamma^{\mathbf{k}}, \triangle\pi^{\mathbf{k}}, \triangle z^{\mathbf{k}})$ is calculated in which the next step is taken. However, the new solution should remain inside the feasible region, i.e. $\gamma^{\mathbf{k}} > 0$ and $z^{\mathbf{k}} > 0$. This is ensured through a suitable choice of the step length parameter $\alpha^{\mathbf{k}}$ (Step 2). In particular, maximum possible primal step length $\alpha^{\mathbf{k}}_{\gamma,max}$ and dual step length $\alpha^{\mathbf{k}}_{z,max}$ are found via the following ratio tests [Kwon, 2013, p. 235]:

$$\alpha^{\mathbf{k}}_{\gamma,max} = \min\left(1, \min_{j:\Delta\gamma^{\mathbf{k}}_j<0} -\frac{\gamma^{\mathbf{k-1}}_j}{\Delta\gamma^{\mathbf{k}}_j}\right);\ \alpha^{\mathbf{k}}_{z,max} = \min\left(1, \min_{j:\Delta z^{\mathbf{k}}_j<0} -\frac{z^{\mathbf{k-1}}_j}{\Delta z^{\mathbf{k}}_j}\right). \tag{3.37}$$

In practical implementations of interior-point methods these step lengths are *damped*, i.e. shortened, by parameter $\eta^{\mathbf{k}} \in (0,1)$ ensuring an adequate distance of $\gamma^{\mathbf{k}}$ and $z^{\mathbf{k}}$ to 0. For instance, Ferris et al. [2007, p. 211] set $\eta^{\mathbf{k}} = \max\{0.9995, 1-\tau^{\mathbf{k}}\}$ in each iteration. As the optimal solution is approached, i.e. $\tau^{\mathbf{k}} \to 0$, no damping is necessary, so $\eta^{\mathbf{k}} \to 1$.

The new solution is then computed (Step 2) as

$$\begin{aligned} \gamma^{\mathbf{k}} &= \gamma^{\mathbf{k-1}} + \alpha^{\mathbf{k}}_{\gamma}\Delta\gamma^{\mathbf{k}}, \\ (\pi^{\mathbf{k}}, z^{\mathbf{k}}) &= (\pi^{\mathbf{k-1}}, z^{\mathbf{k-1}}) + \alpha^{\mathbf{k}}_{z}(\Delta\pi^{\mathbf{k}}, \Delta z^{\mathbf{k}}), \end{aligned} \tag{3.38}$$

with $\alpha^{\mathbf{k}}_{\gamma} = \min\{1, \eta^{\mathbf{k}}\,\alpha^{\mathbf{k}}_{\gamma,max}\}$ and $\alpha^{\mathbf{k}}_{z} = \min\{1, \eta^{\mathbf{k}}\,\alpha^{\mathbf{k}}_{z,max}\}$.

The interior-point algorithm can be terminated (Step 3) if the relative norm of the primal and dual residuals as well as complementarity slackness is smaller than a predefined tolerance level ϵ:

$$\begin{aligned} \frac{\|\tilde{t} - W\gamma^{\mathbf{k}}\|}{1 + \|\tilde{t}\|} &\le \epsilon \\ \frac{\|f - W^T\pi^{\mathbf{k}} - z^{\mathbf{k}}\|}{1 + \|f\|} &\le \epsilon \\ \frac{|f^T\gamma^{\mathbf{k}} - \tilde{t}^T\pi^{\mathbf{k}}|}{1 + |f^T\gamma^{\mathbf{k}}|} &\le \epsilon; \end{aligned} \tag{3.39}$$

see for example Wright [1997, p. 226].[38] The relative residuals are preferred termination criteria as they allow to measure the absolute residuals (nominators in (3.39)) in relation to the norm of the right-hand sides $\tilde{t}$ and f, respectively.

Efficient Determination of Search Directions

The most expensive step within interior-point methods is the determination of the search direction $(\Delta\gamma^{\mathbf{k}}, \Delta\pi^{\mathbf{k}}, \Delta z^{\mathbf{k}})$ by solving (3.35) in each iteration $\mathbf{k}$. Now, the idea of the specialized interior-point method (SIPM) is to exploit the specific structure of the recourse matrix W in (3.5) in order to reduce the computational effort. To simplify the presentation, the iteration counter $\mathbf{k}$ is omitted for now.

[38] See Appendix A.3 for the proof of convergence.

In practical implementations of interior-point methods, $\triangle z$ is eliminated from (3.35) by setting

$$\triangle z = \Gamma^{-1}(\mu e - \Gamma Z e - Z \triangle \gamma), \tag{3.40}$$

leading to the so-called *saddle point system*[39]:

$$\underbrace{\begin{bmatrix} -D & W^T \\ W & 0 \end{bmatrix}}_{\mathcal{S}} \begin{bmatrix} \triangle \gamma \\ \triangle \pi \end{bmatrix} = \begin{bmatrix} r_d \\ r_p \end{bmatrix}, \tag{3.41}$$

with saddle point matrix $\mathcal{S} \in \mathbb{R}^{(n_2+m_2)\times(n_2+m_2)}$, diagonal matrix D

$$D = \Gamma^{-1} Z = \begin{bmatrix} \frac{z_1}{\gamma_1} & \dots & 0 \\ \vdots & \ddots & \vdots \\ 0 & \dots & \frac{z_{n_2}}{\gamma_{n_2}} \end{bmatrix} \in \mathbb{R}^{n_2 \times n_2},$$

and right hand sides

$$\begin{aligned} r_d &= f - W^T \pi - \mu \Gamma^{-1} e \\ r_p &= \tilde{t} - W \gamma. \end{aligned}$$

One can further reduce the saddle point system in (3.41) by eliminating $\triangle \gamma$:

$$\triangle \gamma = -D^{-1}(r_d - W^T \triangle \pi), \tag{3.42}$$

such that

$$\begin{aligned} & W(-D^{-1}(r_d - W^T \triangle \pi)) = r_p \\ \Longleftrightarrow\ & W D^{-1} W^T \triangle \pi - W D^{-1} r_d = r_p. \end{aligned}$$

Therefore, (3.41) is reduced to the system of *normal equations*:

$$W D^{-1} W^T \triangle \pi = rhs, \tag{3.43}$$

where $rhs = r_p + W D^{-1} r_d$. The matrix $W D^{-1} W^T$ is called *Schur complement*[40] of the saddle point matrix $\mathcal{S}$. Since the diagonal matrix D has only positive entries on its

[39] The term *reduced KKT system* is also often used in the literature.

[40] Given a block matrix $X = \begin{bmatrix} A & B \\ C & D \end{bmatrix} \in \mathbb{R}^{(n+m)\times(n+m)}$ with $A \in \mathbb{R}^{n\times n}$, $B \in \mathbb{R}^{n\times m}$, $C \in \mathbb{R}^{m\times n}$ and $D \in \mathbb{R}^{m\times m}$ and assuming the inverse of A exists, then the Schur complement of A in X is defined as $S = D - CA^{-1}B$ [Boyd and Vandenberghe, 2004, p. 650].

diagonal, it is positive definite[41]. In addition, the matrix $WD^{-1}W^T$ is positive definite as W has full row rank, i.e. all row vectors are linearly independent.[42]

The general procedure of today's interior-point solvers is to factorize $WD^{-1}W^T$ (3.43) in order to simplify the calculation of $\Delta\pi$ [Gondzio, 2012]. If dense columns exist in W, as it is often the case in practice, the matrix $WD^{-1}W^T$ is also dense. Factorizing such dense matrices is generally expensive and several techniques are proposed in the literature for dealing with dense columns in the constraint matrix [Zhang, 1998].

However, in the case of two-stage stochastic programs the recourse matrix W in (3.5) possesses a block triangular structure, leading to a sparse structured Schur complement $WD^{-1}W^T$. In particular, $WD^{-1}W^T$ is a *block arrowhead matrix* for W as defined in (3.5):

$$WD^{-1}W^T = \begin{bmatrix} u^T & & & \\ & \ddots & & \\ & & u^T & \\ H_1 & \cdots & H_{|R|} & Q \end{bmatrix} \begin{bmatrix} D_1^{-1} & & & \\ & \ddots & & \\ & & D_R^{-1} & \\ & & & D_Q^{-1} \end{bmatrix} \begin{bmatrix} u & & H_1^T \\ & \ddots & \vdots \\ & & u & H_{|R|}^T \\ & & & Q^T \end{bmatrix}$$

$$= \begin{bmatrix} u^T D_1^{-1} u & & & u^T D_1^{-1} H_1^T \\ & \ddots & & \vdots \\ & & u^T D_{|R|}^{-1} u & u^T D_{|R|}^{-1} H_{|R|}^T \\ H_1 D_1^{-1} u & \cdots & H_{|R|} D_{|R|}^{-1} u & \sum_{r=1}^{|R|} H_r D_r^{-1} H_r^T + Q D_Q^{-1} Q^T \end{bmatrix},$$

where entries only occur on the diagonal, in the last block row and last block column.[43] The pattern of the block arrowhead matrix allows the definition of block matrices in the following way:

$$WD^{-1}W^T = \begin{bmatrix} A & C^T \\ C & B \end{bmatrix} \in \mathbb{R}^{(|R|+|N|\cdot|K|)\times(|R|+|N|\cdot|K|)}, \tag{3.44}$$

with a diagonal matrix A:

$$A = \begin{bmatrix} u^T D_1^{-1} u & & \\ & \ddots & \\ & & u^T D_{|R|}^{-1} u \end{bmatrix} \in \mathbb{R}^{|R|\times|R|},$$

$$C = \begin{bmatrix} H_1 D_1^{-1} u & \cdots & H_{|R|} D_{|R|}^{-1} u \end{bmatrix} \in \mathbb{R}^{|N|\cdot|K|\times|R|},$$

[41] A matrix $A \in \mathbb{R}^{n\times n}$ is positive definite if for any vector $\nu \in \mathbb{R}^n \setminus \{0\}$ the following holds: $\nu^T A \nu > 0$ [Golub and Van Loan, 2013, p. 159].

[42] If a matrix $M \in \mathbb{R}^{n\times n}$ is positive definite and matrix $U \in \mathbb{R}^{m\times n}$ with $m \leq n$ has full rank, i.e. $rank(U) = m$, then UMU^T is positive definite [Horn and Johnson, 2012, p. 399].

[43] From now on the analysis is based on the specific recourse matrix W in (3.5) for the Rawls and Turnquist [2010] model. As discussed in Section 3.3.2, the matrix ordering algorithm of Hu et al. [2000] can arrange W to a block triangular structure such that $WD^{-1}W^T$ is a block arrowhead matrix.

and

$$B = \sum_{r=1}^{|R|} H_r D_r^{-1} H_r^T + Q D_Q^{-1} Q^T \in \mathbb{R}^{|N|\cdot|K| \times |N|\cdot|K|}.$$

Then the system of normal equations (3.43) can be rewritten as

$$\begin{bmatrix} A & C^T \\ C & B \end{bmatrix} \begin{bmatrix} \triangle\pi_1 \\ \triangle\pi_2 \end{bmatrix} = \begin{bmatrix} rhs_1 \\ rhs_2 \end{bmatrix}, \tag{3.45}$$

with a corresponding partitioning of $\triangle\pi$ and the right hand side in (3.43). Finally, eliminating $\triangle\pi_1$ from (3.45)

$$\triangle\pi_1 = A^{-1}(rhs_1 - C^T \triangle\pi_2), \tag{3.46}$$

leads to the equation:

$$\tilde{\mathcal{S}} \triangle\pi_2 = rhs_2 - CA^{-1} rhs_1, \tag{3.47}$$

where $\tilde{\mathcal{S}} = (B - CA^{-1}C^T) \in \mathbb{R}^{|N|\cdot|K| \times |N|\cdot|K|}$ is symmetric and positive definite and represents the Schur complement of $WD^{-1}W^T$ in (3.44). The Schur complement of a (2,2)-block matrix $T = \begin{bmatrix} A & C^T \\ C & B \end{bmatrix} \in \mathbb{R}^{t \times t}$ is symmetric and positive definite if T is itself symmetric and positive definite and A is positive definite [Castro, 2000]. The (1,1)-block matrix A in (3.44) is a positive definite diagonal matrix since it is a product of vector u and diagonal matrix D each with solely positive entries. As $WD^{-1}W^T$ in (3.44) is symmetric and positive definite (see page 66) the Schur complement $\tilde{\mathcal{S}} = (B - CA^{-1}C^T)$ is therefore symmetric and positive definite.

After obtaining $\triangle\pi_2$ from (3.47), one can determine the complete search direction $(\triangle\gamma, \triangle\pi, \triangle z)$ just by back-solving the remaining equations (3.46), (3.42) and (3.40). Note that these equations just require the inverses of diagonal matrices A, D and Γ, which are computationally inexpensive. It should be mentioned that A is a diagonal matrix for the Rawls and Turnquist [2010] case since u^T is a vector in this model. If u is a matrix, A would be block diagonal. In this case, (3.46) can be solved for each diagonal block $u^T D_i^{-1} u$ for $i = 1, ..., |R|$ instead of computing the inverse of A directly.

Now, the most expensive task is to solve (3.47) efficiently. Since $\tilde{\mathcal{S}}$ is symmetric and positive definite the *Cholesky factorization* [Golub and Van Loan, 2013, p. 163] can be used, leading to

$$\tilde{\mathcal{S}} = \mathcal{L}\mathcal{L}^T,$$

where $\mathcal{L} \in \mathbb{R}^{|N|\cdot|K| \times |N|\cdot|K|}$ is a lower triangular matrix with positive entries on its diagonal.

With such a factorization of $\tilde{\mathcal{S}}$, equation (3.47) can be solved through simple forward and backward substitutions:

$$\mathcal{L}y = rhs_2 - CA^{-1} rhs_1$$
$$\mathcal{L}^T \triangle\pi_2 = y.$$

Castro [2000] proposes a similar solution approach for multi-commodity network problems where $WD^{-1}W^T$ is also a block arrowhead matrix. However, W does not represent a recourse matrix in his application since the underlying problem is not a two-stage stochastic program. In addition, the Schur complement $\tilde{\mathcal{S}}$ may reach a high degree of density and depends on the number of arcs which is very large in the multi-commodity network problem considered by Castro [2000]. Therefore, the author suggests an iterative algorithm, namely the preconditioned conjugate gradient method, instead of an exact method to solve (3.47). In contrast, the matrix $\tilde{\mathcal{S}}$ in (3.47) for the Rawls and Turnquist [2010] model has a size of $|N| \cdot |K| \times |N| \cdot |K|$ which is generally small. For instance, natural disasters like storms or floods hit specific regions, i.e. the number of potential locations $|N|$ may not exceed a few hundreds. Moreover, the range of relief items $|K|$ is limited to the basic necessities. As a result, the product $|N| \cdot |K|$ is expected to be of manageable size in humanitarian applications and therefore the Cholesky factorization represents a reasonable choice.

A well-known drawback of interior-point methods arises when the optimal solution for the original problem (3.18) and (3.16) is approached, i.e. $\mu \to 0$. In this case either the primal variable $z_j \to 0$ or the dual slack variable $\gamma_j \to 0$ for each $j = 1, ..., n_2$ due to the complementarity conditions in (3.33). As a consequence the diagonal elements z_j/γ_j $(j = 1, ..., n_2)$ in D tend to 0 or infinity. Therefore, matrices containing D, namely S in (3.41), $WD^{-1}W^T$ in (3.44) and $\tilde{\mathcal{S}}$ in (3.47), may become highly ill-conditioned in the final iterations. Ill-conditioned matrices are characterized by a large condition number. In general, the condition number of a matrix $A \in \mathbb{R}^{n \times n}$ reflects the sensitivity of a linear system $Ax = b$, measuring the effects on the solution $x \in \mathbb{R}^n$ for small changes in A and the right-hand side $b \in \mathbb{R}^n$. For a large condition number small perturbations, e.g. due to round-off errors, may cause severe errors in the solution.

Unfortunately, applying Cholesky factorization to such an ill-conditioned matrix may provide highly inaccurate solutions or the Cholesky factorization may even break down [Zhang, 1998]. However, in the case of two-stage stochastic programming, valid optimality cuts can be determined even if the sub-problems are not solved to a high degree of accuracy [Rahmaniani et al., 2017].

In this section it was shown how the specialized interior-point method SIPM can exploit the block triangular structure of W in (3.5) and therefore the block arrow head structure of $WD^{-1}W^T$ in (3.44), leading to a simplified determination of search directions. Instead of solving the saddle point system (3.41) with
$S \in \mathbb{R}^{(|R|\cdot|K|+3|N|\cdot|K|+2|R|)\times(|R|\cdot|K|+3|N|\cdot|K|+2|R|)}$, the search direction is based on the much smaller system (3.47) with $\tilde{\mathcal{S}} \in \mathbb{R}^{|N|\cdot|K|\times|N|\cdot|K|}$. Using SIPM for solving the sub-problems leads to the accelerated version of the L-shaped method. In order to validate the advantage of the theoretically developed accelerated L-shaped method with respect to solution times, the results of numerical experiments are presented in the following.

4 Numerical Experiments

This chapter is dedicated to showing the benefits of the accelerated L-shaped method with respect to fast solution times for large-scale problems. The primary goal is to verify the efficiency of the specialized interior-point method SIPM, introduced in Section 3.3.3, independent of the underlying problem size. For this purpose three case studies, namely a small-, medium- and large-scale case study, are presented. The large-scale case study is based on the one proposed by Rawls and Turnquist [2010], but more realistic data and sources are used. Section 4.1.1 provides the details on the case study design. After specifying how the numerical tests are carried out (Section 4.1.2), the performance of different solution methods for the three case studies are evaluated (Section 4.1.3). In particular, the commercial solver Gurobi, the standard L-shaped method and its accelerated version are compared with respect to their solution times. The main purpose of Section 4.1.3 is to verify the usefulness of the accelerated L-shaped method especially for larger problems arising in realistic settings. In addition, location decisions are evaluated against the solution suggested by Rawls and Turnquist [2010].

In order to analyze how solution times may change when input data are changed, the large-scale case study is modified in Section 4.2.1. The three-day forecast for hurricane Katrina in 2005 is used to adjust the scenario-dependent parameters to this specific threat. The resulting case study has the same size as the one proposed in Section 4.1.1 but is adapted to the available hurricane forecast. The aim of Section 4.2.2 is to demonstrate that the good performance of SIPM remains unaffected by data changes.

Future research directions for two-stage stochastic programs where the second-stage problem is even larger than in the newly proposed case study are outlined in Section 4.3.

4.1 Realistic Large-Scale Case Study

In this section[1] a realistic case study is designed to generate realistic location and pre-positioning decisions and to test the efficiency of the accelerated L-shaped method with respect to solution times for large-scale problems. For reasons of comparability, this case study is based on the one used in Rawls and Turnquist [2010] considering the hurricane-prone Gulf and Atlantic coast. As this region is especially prone to hurricanes, pre-positioning of relief items for the annual hurricane season is a reasonable preparation measure. This is done by many humanitarian agencies like Direct Relief. In this case study, decisions have to be made where to open relief facilities on America's southeast coast and

[1]Section 4.1 is partially taken from Grass et al. [2018].The data presented in this section is partially taken from Rams [2016].

E. Graß, *An Accelerated Solution Method for Two-Stage Stochastic Models in Disaster Management*, Mathematische Optimierung und Wirtschaftsmathematik | Mathematical Optimization and Economathematics,
https://doi.org/10.1007/978-3-658-24081-3_4

how much to store there in order to be prepared for future hurricanes. However, such decisions are difficult due to the highly uncertain nature of hurricanes. Although storms and hurricanes are predictable in the short run, general forecasts for the annual season are hardly possible. Therefore, historical hurricanes, dating back to 1900, are used to define a multitude of more realistic scenarios which may help aid agencies in their decision-making process.

According to the Saffir-Simpson scale hurricanes are classified into five categories depending on their wind speed given in Table 4.1 [NHC, 2017b]. For instance, a category 3 hurricane reaches wind speeds between 178 km/h and 208 km/h.

Table 4.1: Hurricane categories [NHC, 2017b]

Category	Wind Speed	
1	119-153 km/h	74-95 mph
2	154-177 km/h	96-110 mph
3	178-208 km/h	111-129 mph
4	209-251 km/h	130-156 mph
5	≥252 km/h	≥156 mph

The American database *International Best Track Archive for Climate Stewardship* (IBTrACS) of the *National Centers for Environmental Information* has completely registered all storms and hurricanes since 1900. In this time frame, the landfall of 203 hurricanes can be identified [URI and GSO, 2016] with an average of nearly two hurricanes per year [AOML, 2016]. Figure 4.1 shows areas which were hit by hurricanes of different categories from 1900-2014. Note that the highest category is given in this figure. Especially three regions (dark violet) were hit by hurricanes of the highest category 5 in the past. One can deduce that these areas were also affected by weaker hurricanes and are therefore under particular threat. In contrast, counties far away from the coast were hit only by weaker hurricanes (light pink areas) or by storms which are not classified as hurricanes, i.e. storms with wind speeds of less than 119 km/h (green areas). In total, 900 counties on the American southeast coast have been hit by storms or hurricanes in the past.

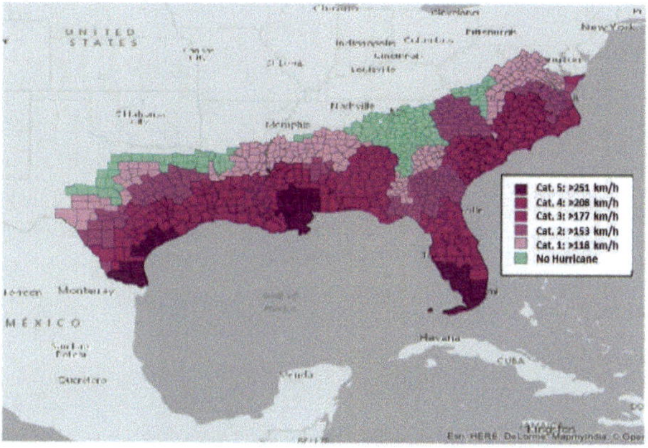

Figure 4.1: Counties hit by hurricanes of different categories from 1900-2014 (based on IB-TrACS)

4.1.1 Data

The case study proposed here is based on the one given in Rawls and Turnquist [2010] but supplemented by additional and more realistic data, leading to an increased problem size. In the following, the way parameter values[2] are determined for the large-scale case study are described in more detail. In particular, sources used to obtain more realistic data are given.

Storage Facilities

For the large-scale case study the 900 vulnerable counties shown in Figure 4.1 are grouped into 90 nodes, where each node represents the geographical center of the adjacent counties; see Figure 4.2. In contrast, the network considered in Rawls and Turnquist [2010] contains only 30 nodes (see Figure 2.7 in Section 2.3). The increased number of nodes allows for a more detailed analysis since the area covered by each node and the distances between them are reduced. This leads to more precise estimations of affected people and the differentiation between areas which are actually hit by a hurricane and those which are close but less affected.

[2] All parameter values for the large-scale case study can be found in the Excel file *LargeCaseStudy.xlsx* in the folder "DataImport" on https://zenodo.org/ [Grass, 2017].

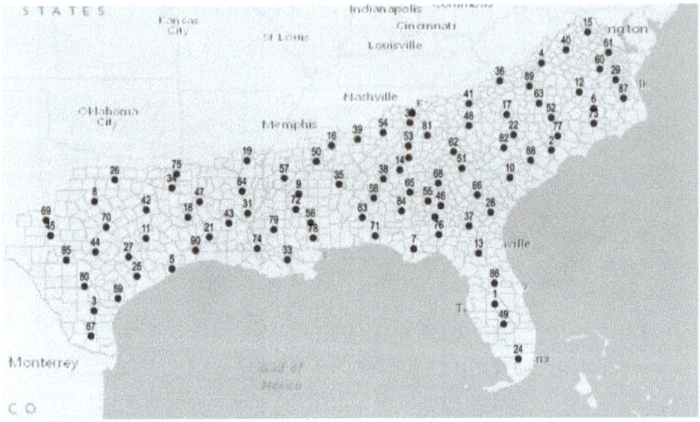

Figure 4.2: Network containing 90 nodes on the Gulf and Atlantic coast

Each of these nodes can be hit by hurricanes, leading to demand for relief items. In order to provide first-aid supplies to the affected population, storage facilities can be opened at these nodes even before a hurricane strikes. As indicated by Table 4.2, three types of facility categories are available, namely small-, medium- and large-sized facilities. For comparability reasons, corresponding opening costs F_l as well as capacities M_l of these facilities are taken from Rawls and Turnquist [2010].

Table 4.2: Opening costs and storage capacities [Rawls and Turnquist, 2010]

Size category l	Fixed costs F_l (in US$)	Capacity M_l (in ft^3)
1 small	19,600	36,400
2 medium	188,400	408,200
3 large	300,000	780,000

For instance, a medium-sized facility has a storage capacity of 408,200 ft^3 with fixed costs of US$ 188,400. Information on how these values are determined is not provided by Rawls and Turnquist [2010]. Indeed, gathering such information is difficult in practice since relief organizations are faced with problems in reliably estimating opening costs. In contrast, calculating costs concerning the procurement, storage and transportation of relief items is much easier due to existing contracts with, e.g., suppliers as discussed next.

Specifications of Relief Items

In Rawls and Turnquist [2010], water, meals ready-to-eat (MREs) and medical sets are considered as relief items. For the large-scale case study two types of relief items are added, namely tents and first-aid kits (FAKs). Water, tents and MREs belong to the basic equipment of evacuated persons. Originally used for soldiers during combat and battle operations, MREs are nowadays also distributed to civilians after natural disasters. Each meal has on average 1200 kcal [RCRC, 2009, p. 67].

Medical sets contain essential medicines for preventing epidemics and treating chronically ill persons. Chemicals for water purification are also included in the medical sets, ensuring safe drinking water. FAKs are used for people with, e.g., serious injuries and heart attacks and should be available as quickly as possible in the aftermath of a disaster. Therefore, the aid agency Direct Relief stores such medical equipment and drugs like antibiotics and allergy pills in their partner hospitals during the entire hurricane season.

Depending on the hurricane's category, open facilities where relief items are pre-positioned can be destroyed to varying extents. The proportion of relief items which is not destroyed by the hurricane, i.e. $\rho_i^{k,s}$ defined in Section 2.3.1, can be used in the aftermath to help those in need. Similar to Rawls and Turnquist [2010], it is assumed that hurricanes of category 4 and 5 cause total destruction of storage facilities, leading to $\rho_i^{k,s} = 0$. In the case of a category 3 hurricane, 50% of relief items are still usable ($\rho_i^{k,s} = 0.5$) and 100% for minor hurricanes of category 1 and 2 ($\rho_i^{k,s} = 1$).

The specifications of different types of relief items are summarized in Table 4.3. According to MSF [2016], purchase prices q^k are negotiated with the corresponding supplier in advance, avoiding price increases in emergency situations. The specific purchase price q^k, the transport weight u^k (measured in kg) and space requirement b^k (measured in ft^3) for each type of relief item k are based on the *South Carolina Emergency Operations Plan* [SCEMD, 2007] as well as the *items catalogue* of IFRC [2016]. For instance, one tent has a size of about 9.95 ft^3 and weights 55 kg. In contrast, Rawls and Turnquist [2010] state that their chosen cost rates and space requirements can be seen as realistic but they do not provide a specific source. In addition, values for weights, i.e. u^k, are not specified.

Table 4.3: Purchase price, weight and space requirement of each relief item (based on IFRC [2016] and SCEMD [2007])

Item k	Unit	Purchase price q^k (in US$/unit)	Weight u^k (in kg/unit)	Space b^k (in ft^3/unit)
1 Water	one liter	0.68	1.13	0.093
2 MRE	one package	3.36	0.76	0.1029
3 Tent	one piece	310	55	9.9499
4 Medical Set	one box	770	45	6.3566
5 FAK	one box	870	10.5	2.2434

It is assumed that standard 48ft trucks are used with a load capacity of 20,000 kg to deliver pre-positioned relief items on roads.[3] Based on MSF [2016], costs for transportation $c_{i,j}$ are set to US$ 0.50 per kg and increase linearly with the respective distance. Distances

[3]Note that the Rawls and Turnquist [2010] model does not distinguish between different modes of transport. Although air transport is relatively common, delivering aid supplies on roads is most frequently used and therefore considered in this case study.

between the corresponding nodes are determined by the ArcGIS software where main roads and highways are considered, resulting in 8,100 arcs. Relief items can be transported only in the case of undamaged arcs, i.e. if arc capacity $U^s_{(i,j)}$ is large enough for a specific scenario s. Here, $U^s_{(i,j)}$ =100,000 kg should be sufficient and is used for hurricane category 1-3 and for all arcs (i, j). It is assumed that major hurricanes of category 4 and 5 cause complete destruction, leading to $U^s_{(i,j)} = 0$ for particular arcs (i, j) and scenario s. This is a reasonable choice since fallen trees and floods can render roads impassable. Again, concrete values for $U^s_{(i,j)}$ are not given in Rawls and Turnquist [2010].

The last type of cost rates associated with relief items are penalty costs for unused items h^k and unsatisfied demand p^k. In order to have better comparability of the numerical results in Section 4.1.3 these cost rates are based on the ones specified by Rawls and Turnquist [2010]. In particular, penalty costs for unused items h^k are one quarter of the purchase costs q^k and penalty costs for unsatisfied demand p^k are set to $p^k = 10 \cdot q^k$. Penalizing unsatisfied demand much more than surplus is reasonable since first-aid supplies are essential to human life in the event of an emergency. As it is almost impossible to assess the consequences of insufficient quantities of relief items with respect to human lives, these penalty costs are purely fictitious.

Scenario Generation

Actual effects of hurricanes on the network structure and demand are uncertain and have to be captured through the use of scenarios. As already indicated in Section 2.1.2, designing scenarios is not an easy task. Location, type and scale of a disaster have to be taken into account with the respective probability of occurrence. For instance, Rawls and Turnquist [2010] use ten historical hurricanes of category 3-5 and five historical hurricanes of category 1-2 to design 51 scenarios. Figure 4.3 illustrates which nodes in the network are assumed to be affected by these hurricanes.

When comparing hurricane-prone regions defined in Figure 4.1 and Figure 4.3 it is noticeable that, in contrast to the newly proposed case study, the western area around Brownsville and Corpus Christi (node 1 and 2 in Figure 4.3) as well as areas further inland are not considered as vulnerable in the case study of Rawls and Turnquist [2010]. This fact is reflected by their scenario definitions where demand and network destruction can only arise at the labeled nodes in Figure 4.3. Consequently, the resulting location decisions depend on these scenario definitions, as will be shown in Section 4.1.3.

For more realistic scenarios the IBTrACS database is used providing frequencies, wind speeds and the paths of each hurricane since 1900. Based on this information, counties hit by historical hurricanes of different categories can be identified (see Figure 4.1). The width of hurricanes, namely the strike circle, may vary extremely. In 2004, category 4 hurricane *Charley* had a strike circle of 16-24 km whereas one year later *Katrina* (category 5 in the meantime) exceeded 200 km. For this case study, a diameter of 161 km on average is selected such that counties located within 80.5 km of a hurricane path are assumed to be affected.

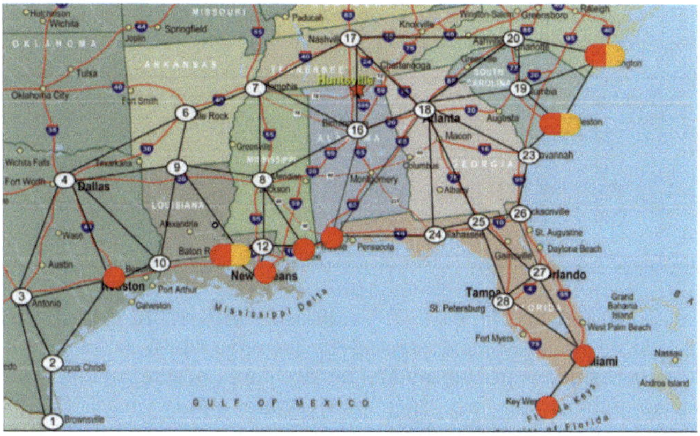

Figure 4.3: Landfall nodes of major (red circles) and minor (yellow circles) hurricanes in Rawls and Turnquist [2010]

For the scenario definitions, all 203 historical hurricanes are analyzed and those with similar characteristics, e.g. category and path, are clustered together. As a result, 118 scenarios are defined consisting of up to seven similar hurricanes. Assuming that each of these hurricanes is equally likely to occur the scenario probability P_s is the sum of these probabilities. For instance, a scenario consisting of three hurricanes has a probability of $P_s = 3/203$.

Each scenario contains the information on demand $v_i^{k,s}$ and network destruction, i.e. $U_{(i,j)}^s$ and $\rho_i^{k,s}$. As described above, depending on the hurricane location and category defined in each scenario, certain roads and facilities can be destroyed, leading to $U_{(i,j)}^s = 0$ and $\rho_i^{k,s} = 0$ or $\rho_i^{k,s} = 0.5$ for corresponding arcs (i, j) and nodes i, respectively.

In addition, demand for relief items $v_i^{k,s}$ is also influenced by the scenario definitions as described next.

Demand

A variety of factors like hurricane category, floods or degree of injuries can have an effect on demand for specific relief items. Here, it is assumed that the damage potential of a hurricane, as defined in Table 4.4, influences the actual demand for each scenario.

For instance, 5% of the population in affected counties have to be supplied with pre-positioned relief items in the event of a category 3 hurricane [NOAA and AOML, 2016]. The population size of each county in 2016 is taken from the Census Bureau [CB, 2016] and added up for all counties belonging to the same node.

Table 4.4: Percentage of population to be provided with relief items depending on the damage potential of hurricanes [NOAA and AOML, 2016]

Category	Wind Speed (km/h)	Damage Potential	Percentage of affected population (%)
1	119 - 153	1	0.1
2	154 - 177	10	1
3	178 - 208	50	5
4	209 - 251	250	25
5	$\geq$252	500	50

Aid items should last for two days after a disastrous event. In this time frame relief organizations can determine demand more exactly and order additional supplies if necessary. In Table 4.5 required items per person and for two days are summarized. Following the emergency plan of South Carolina, three liters of water and two meals (MREs) per day and person are required [SCEMD, 2007], i.e. six liters of water and four MREs for each person have to be pre-positioned for the entire period of two days (see Table 4.5).

Table 4.5: Required relief items per person for two days

No.	k	Unit	Required items pp (2 days)
1	Water	one liter	6
2	MRE	one package	4
3	Tent	one piece	1/5
4	Medical Set	one box	1/1,000
5	FAK	one box	1/5

As a tent has space for up to five persons only one fifth of a tent per person is required. In order to prevent the outbreak and spread of diseases, vaccines have to be distributed to everyone in affected regions. One box of medical supplies can provide 1,000 people with medication against epidemics. Similarly, one box of first-aid kits can supply 5 people with bandages for external injuries and resuscitation equipment for heart attacks.

As for other parameters, Rawls and Turnquist [2010] do not refer to how demand is defined. In summary, Table 4.6 compares sources used for designing the large-scale case study and sources given in Rawls and Turnquist [2010]. Except for the indication of past hurricanes taken from the *Atlantic Oceanographic and Meteorological Laboratory* (AOML), no further sources are specified by Rawls and Turnquist [2010].

Table 4.6: Data sources of the case study proposed by Rawls and Turnquist [2010] and the newly proposed case study

	Source	
Data	Large-Scale Case Study	Rawls and Turnquist [2010]
Network	ArcGIS	No source specified
Population size	U.S. Census Bureau 2016	No source specified
Cost determination	IFRC; MSF	No source specified
Scenario generation	IBTrACS	AOML
Demand	SCEMD [2007]; NOAA and AOML [2016]	No source specified

Resulting Problem Size

The resulting large-scale case study has a problem size of 4,885,920 decision variables and 1,009,080 constraints. To be more specific, the number of first-stage decision variables for the large-scale case study is 720, of which 270 are binary, whereas second-stage decision variables amount to 4,885,200. Due to such an immense difference in size, different requirements for solving the first- and second-stage problem exist when applying the L-shaped method. As already discussed in Section 3.3.1, solving the first-stage problem does not represent a computational burden, whereas an efficient solution method is needed for the second-stage problem with almost five million decision variables. In particular, the second stage consists of 118 sub-problems (equivalent to the number of scenarios), where each of these has 41,400 decision variables. In every iteration of the accelerated L-shaped method the primal (3.18) and dual sub-problem (3.16) are solved via the specialized interior-point method SIPM for each scenario s. The corresponding recourse matrix W in (3.5) for the large-scale case study has a size of 8,550 $\times$ 49,500 (including slack variables $g_{(i,j)}$ introduced on page 48) and a sparsity pattern shown in Figure 4.4. The figure illustrates the lower block triangular structure of W where the number of non-zero entries is 129,600, representing just 0.03% of the entire matrix. In other words, the recourse matrix W in (3.5) is highly sparse, which is a beneficial property especially for iterative algorithms.

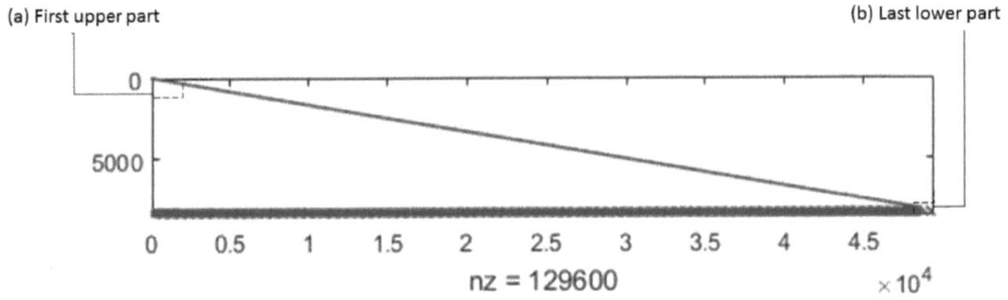

Figure 4.4: Sparsity pattern of the recourse matrix W in (3.5) where nz denotes the number of non-zero entries

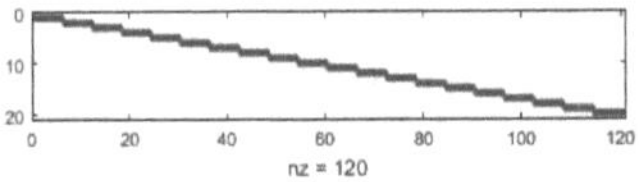

(a) First upper part

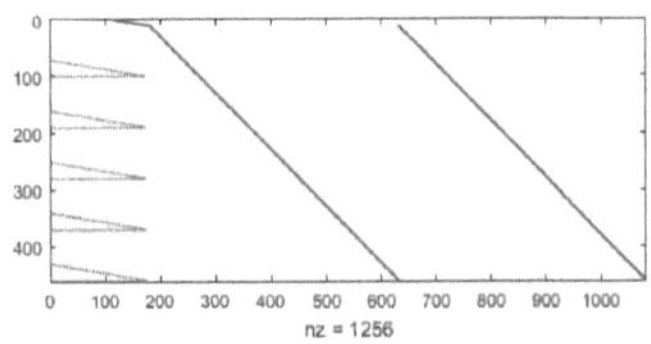

(b) Last lower part

Figure 4.5: Enlarged extracts of the recourse matrix W showing the first upper part (a) and last lower part (b) in Figure 4.4

When zooming in the first upper part of W (Figure 4.5a), the block diagonal consists of row vectors $u^T = [u^1\ u^2\ u^3\ u^4\ u^5\ 1]$. The last lower part of W is illustrated in Figure (4.5b), where matrix Q (represented by two diagonals) and a part of H are shown.[4]

Due to the block triangular structure of W in (3.5) and Figure 4.4, the Schur complement $WDW^T = \begin{bmatrix} A & C^T \\ C & B \end{bmatrix}$ in (3.44) is a block arrowhead matrix. The sparsity pattern of WDW^T for the large-scale case study is as follows:

[4]See page 49 for an example of matrix Q and H.

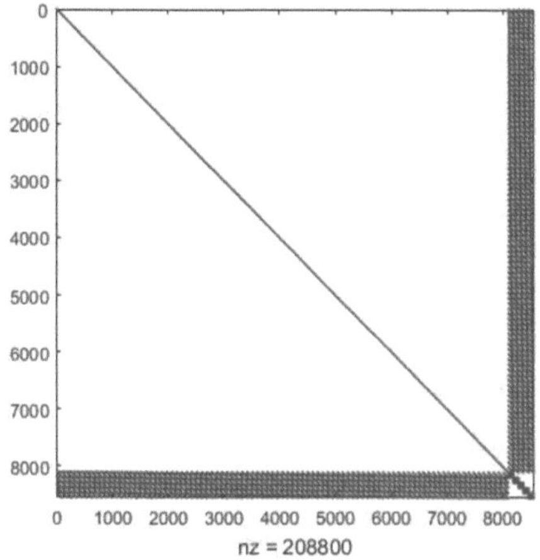

Figure 4.6: Sparsity pattern of the Schur complement $WD^{-1}W^T$ for the large-scale case study

From Figure 4.6 one can see that the (1,1)-block A in (3.44) is a large diagonal matrix, whereas the (2,2)-block $B = \sum_{r=1}^{|R|} H_r D_r^{-1} H_r^T + QD_Q^{-1}Q^T$ is block diagonal and negligibly small in comparison to A. Since the product $QD_Q^{-1}Q^T$ is just a diagonal matrix the block diagonal structure of B results mainly from the zigzag pattern of H (see Figure 4.5b). As discussed in Section 3.2.1 (on page 49), each row in H represents a flow conservation constraint (2.5) for each node $i \in N$ and each relief item of type $k \in K$. Therefore, block diagonal matrix B in Figure 4.6 consists of five blocks, corresponding to the number of relief items $|K|$, where each of these blocks has a size of 90×90, corresponding to the number of nodes $|N|$. Finally, the (2,1)-block C in Figure 4.6 is a rectangular matrix with many more columns than rows. As a result, the final matrix $\tilde{\mathcal{S}} = (B - CA^{-1}C^T)$ in (3.47) has a size of only 450×450 for the newly proposed case study. Note that exploiting the block triangular structure of W in (3.5) has led to a size reduction from $8{,}550 \times 8{,}550$ (matrix $WD^{-1}W^T \in \mathbb{R}^{(|R|+|N|\cdot|K|)\times(|R|+|N|\cdot|K|)}$ in (3.44)) to 450×450 (matrix $\tilde{\mathcal{S}} \in \mathbb{R}^{|N|\cdot|K|\times|N|\cdot|K|}$ in (3.47)). Therefore, factorizing $\tilde{\mathcal{S}}$ via Cholesky can be expected to be less expensive than solving the system of normal equations in (3.43).

Additional Case Studies

In Table 4.7 the sizes of three different case studies, namely a small-, medium- and the large-scale case study, are compared.[5] These case studies are used to illustrate for which problem size the accelerated L-shaped method becomes worthwhile.

Table 4.7: Problem size of a small-, medium- and large-scale case study

Number of	Small	Case Study Medium	Large
Nodes $\|N\|$ and $\|I\|$	30	90	90
Types of relief items $\|K\|$	3	3	5
Facility size categories$\|L\|$	3	3	3
Scenarios $\|S\|$	51	80	118
Constraints	7,608	669,780	1,009,080
Continuous decision variables	147,060	1,987,740	4,885,920
Binary decision variables	90 binary	270 binary	270 binary

The small case study has the same problem size as the one proposed by Rawls and Turnquist [2010] with parameter values taken from their publication. Missing data are completed according to the same principles as for the large-scale case study. For instance, ArcGIS is used to determine the distances between the 30 nodes in Figure 2.7.

In order to define a medium-sized problem, the large-scale case study is roughly halved by reducing the number of item types and scenarios to almost two million decision variables and over 600,000 constraints (see Table 4.7). In particular, the number of item types $|K|$ and the number of scenarios $|S|$ is reduced to 3 and 80, respectively.

Before solving these case studies some technical specifications are given next.

4.1.2 Technical Specifications

All numerical experiments were carried out on a Dell Latitude E7470 laptop with a dual 2.60 GHz Intel Core i7 processor and 8 GB of memory.

All three case studies, summarized in Table 4.7, are solved by the commercial optimization software Gurobi 7.0.0, the standard L-shaped method, introduced in Section 3.2.2, and the accelerated L-shaped method, proposed in Section 3.3.1. Both versions of the L-shaped method are implemented in MATLAB 2016b.[6] The solution procedure is shown schematically in Figure 4.7.

[5] Data for all three case studies can be found in the folder "DataImport" on https://zenodo.org/ [Grass, 2017].

[6] The corresponding MATLAB code `LShaped_MultiCut.m` can be found in Appendix A.4 and in the folder "LShaped" on https://zenodo.org/ [Grass, 2017].

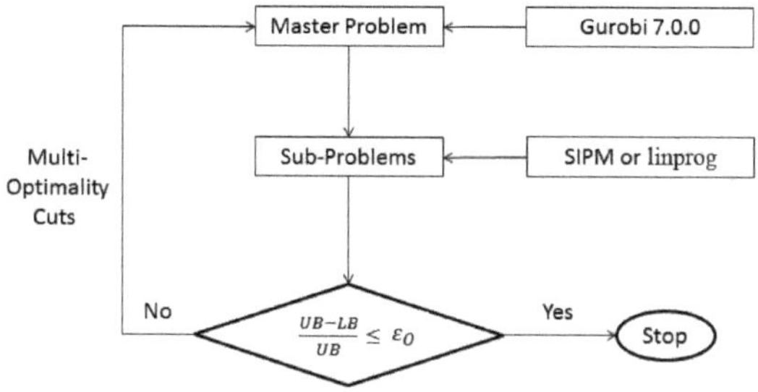

Figure 4.7: Scheme of the standard and accelerated L-shaped method

The standard and accelerated L-shaped method terminate if the relative optimality gap *RelOptGap* in the current iteration[7] is smaller than a predefined overall tolerance level ϵ_O:

$$RelOptGap := \frac{(UB - LB)}{UB} \leq \epsilon_O. \tag{4.1}$$

The lower bound LB is defined as the current objective value of the master problem:

$$LB = \alpha^T \overline{\chi} + \sum_{s=1}^{|S|} \theta_s, \tag{4.2}$$

whereas the upper bound UB is

$$UB = \alpha^T \overline{\chi} + \sum_{s=1}^{|S|} P_s f^T \overline{\gamma}_s. \tag{4.3}$$

As the term $\alpha^T \overline{\chi}$ coincides in (4.2) and (4.3) it remains to compare the approximated second-stage costs $\sum_{s=1}^{|S|} \theta_s$, determined by the multi-optimality cuts in (3.25), and the probability weighted second-stage costs $\sum_{s=1}^{|S|} P_s f^T \overline{\gamma}_s$ for a fixed first-stage solution $\overline{\chi}$. The latter term $\sum_{s=1}^{|S|} P_s f^T \overline{\gamma}_s$ represents the actual probability-weighted second-stage costs when first-stage decisions are already made, whereas in the case of $\sum_{s=1}^{|S|} \theta_s$, first-stage decisions χ can still be chosen.

As long as $RelOptGap > \epsilon_O$, multi-optimality cuts are added to the master problem. The multi-cut approach, as described in Section 3.2.2, is a reasonable choice since the number of scenarios is less than the number of first-stage constraints [Birge and Louveaux, 2011, p. 200].

According to Figure 4.7, the master problem is solved by Gurobi 7.0.0 which is terminated if the relative optimality gap MIPGap is smaller than the tolerance level for the master problem $\epsilon_M = 0.0001$.[8]

[7]Here, the iteration counter ι is omitted to simplify the presentation.

[8]Gurobi 7.0.0 terminates, i.e. considers an optimization problem as solved, if MIPGap=(upper bound−lower bound)/upper bound is smaller than 0.0001 by default [Gurobi, 2017a].

The only difference between the standard and the accelerated L-shaped method in Figure 4.7 is the way the sub-problems are solved. In the case of the standard L-shaped method, MATLAB's built-in function `linprog`[9] is used to solve each sub-problem. In order to reduce solution times for the sub-problems, the specialized interior-point method SIPM (see Section 3.3.3) is applied in the accelerated version of the L-shaped method.[10] Both interior-point methods, i.e. `linprog` and SIPM, are terminated if the sum of relative residuals (first and second term in (4.4)) and the relative optimality gap (third term in (4.4)) is smaller than the tolerance level for the sub-problems $\epsilon_S = 0.00001$:

$$\frac{||\tilde{t} - W\gamma^{\mathbf{k}}||}{\max(1, ||\tilde{t}||)} + \frac{||f - W^T\pi^{\mathbf{k}} - z^{\mathbf{k}}||}{\max(1, ||f||)} + \frac{|f^T\gamma^{\mathbf{k}} - \tilde{t}^T\pi^{\mathbf{k}}|}{\max(1, |f^T\gamma^{\mathbf{k}}|)} \leq \epsilon_S. \tag{4.4}$$

Note that this termination criterion is valid for the function `linprog` but differs from the one defined in (3.39). To ensure comparability of both methods, (4.4) has been adopted for SIPM.

Within SIPM the search direction is based on (3.47) applying the Cholesky factorization. In particular, MATLAB's built-in function `chol` (Cholesky factorization) is used to decompose the matrix $\tilde{\mathcal{S}}$ in (3.47) into a lower triangular matrix $\mathcal{L}$ and its transpose $\mathcal{L}^T$. As described in Section 3.3.3, the Cholesky factorization may become unstable for ill-conditioned matrices at final iterations of interior-point methods. Therefore, SIPM is terminated prematurely if MATLAB issues a warning that the (1,1)-block matrix A in (3.44) is highly ill-conditioned, meaning that one part of diagonal elements is close to 0 and the other part close to a large number. As this drawback of interior-point methods arises only at the end of the process, i.e. if μ approaches 0, the loss of accuracy is negligible. In other words, despite the early termination of SIPM for some scenarios, the achieved accuracy level, i.e. the sum in (4.4), is still close to the tolerance value $\epsilon_S = 0.00001$. Numerical tests showed that if SIPM was terminated prematurely for a specific scenario then the achieved accuracy level was, e.g., 0.000015 and therefore still sufficiently accurate.

4.1.3 Computational Results

The main purpose of this section is to show that the accelerated L-shaped method is particularly valuable for increasing problem sizes and to demonstrate the superiority of SIPM over `linprog` with respect to solution times regardless of the problem size. In addition, results concerning location decisions are evaluated.[11]

[9]The built-in function `linprog` uses a *predictor-corrector primal-dual interior-point* method by default.

[10]The MATLAB code SIPM.m can be found in Appendix A.5 and in the folder "LShaped" on https://zenodo.org/ [Grass, 2017].

[11]All results, i.e. Gurobi's and MATLAB's solution files, for all case studies can be found in the folder "Results" on https://zenodo.org/ [Grass, 2017].

Small-Scale Case Study

Table 4.8 compares the performances of the solution approaches in terms of solution times and accuracies for the small-scale case study. Column "Gurobi" refers to solving the original model in (2.1)-(2.9) directly by Gurobi 7.0.0.[12] In addition, the standard and accelerated L-shaped method, as described in Figure 4.7, are compared. For each of these approaches the required overall solution time for the entire problem ("Overall time") is given together with the relative optimality gap, "MIPGap" for Gurobi and "RelOptGap" in (4.1) for both versions of the L-shaped method. The relative optimality gap MIPGap differs from RelOptGap in (4.1) in the definition of the bounds. For the L-shaped method both bounds are defined in (4.2) and (4.3), respectively, whereas Gurobi's lower bound is defined as the best bound found so far and the upper bound is the incumbent integer solution [Gurobi, 2017b].

In Table 4.8, the average solution time for each sub-problem within the L-shaped method ("Average time per sub-problem") as well as the number of iterations for the L-shaped method ("No. iterations") is shown.

Table 4.8: Performance overview of different solution approaches for the small-scale case study

	Solution Approach		
	Gurobi	L-Shaped	Accelerated L-Shaped
MIPGap/RelOptGap	0.0001	0.0002	0.0005
Overall time	7.8 sec	244.5 sec (4.1 min)	221.3 sec (3.7 min)
Average time per sub-problem		0.07 sec	0.04 sec
No. iterations		29	29

According to Table 4.8, it is not worthwhile to implement the L-shaped method for small problems as Gurobi 7.0.0 solved the small-scale case study within 7.8 seconds with the desired accuracy level of 0.0001. Hence, Gurobi 7.0.0 outperformed both versions of the L-shaped method, which needed 4 minutes and 3.7 minutes, respectively. For small-scale problems solving the master problem in every iteration of the L-shaped method takes much more time than the direct solution of the original problem (2.1)-(2.9). This can be seen in Figure 4.8, where the most time-consuming functions of the L-shaped method are provided by MATLAB's profiler and sorted in descending order.[13] Column "Calls" describes how often a function was executed. For instance, function `intlinprog_gurobi` (or equivalently

[12]The default setting in Gurobi is to use the dual simplex method for the MIP root node and to run the barrier and the simplex method simultaneously on multiple threads for LP problems [Gurobi, 2017].

[13]The complete profile summary can be found in the folder "Results" on https://zenodo.org/ [Grass, 2017].

`gurobi`)[14] is used to solve the master problem and was called 29 times, representing the number of L-shaped iterations. The "Self Time" refers to the time needed for executing a particular function without taking internal sub-functions, so-called *child functions*, into account. Within the L-shaped method (`LShaped_MultiCut`) several sub-functions have to be executed, e.g. the function `gurobi` for solving the master problem and `linprog` for the sub-problems.

Profile Summary
Generated 08-Jul-2017 21:42:53 using performance time.

Function Name	Calls	Total Time	Self Time*
LShaped_MultiCut	1	244.573 s	3.743 s
intlinprog_gurobi	29	136.803 s	0.030 s
gurobi (MEX-file)	29	136.773 s	136.773 s
linprog	1479	103.424 s	0.660 s

Figure 4.8: MATLAB's profiler showing the most time-consuming functions of the standard L-shaped method for the small-scale case study

In total, the standard L-shaped method required 244.573 seconds to solve the small-scale case study (see "Total Time" in Figure 4.8). Here, solving the master problem via `gurobi` was more time-consuming than solving the sub-problems via `linprog`. Overall 1,479 sub-problems[15] (see "Calls" in Figure 4.8) had to be solved, requiring 103.424 seconds. Therefore, 0.07 seconds were needed on average to solve each sub-problem, as stated in Table 4.8. In contrast, the specialized interior-point method SIPM was nearly twice as fast, namely 0.04 seconds. Despite this time reduction, the accelerated L-shaped method is in total only 23 seconds faster than the standard version. Since solving the master problem is more time-consuming than the sub-problems, the time reduction effect of SIPM is almost negligible for small-scale problems.

First-Stage Decisions for the Small-Scale Case Study

In the following, the heuristically[16] found solution of Rawls and Turnquist [2010] is compared to the one of the small-scale case study. According to Rawls and Turnquist [2010] over eight million gallons of water, almost four million packages of MREs and about 19,000 medical sets should be pre-positioned, resulting in procurement costs of about US$ 29.5 million. Similar pre-positioning decisions for the small-scale case study

[14] The function `intlinprog_gurobi` calls the actual solver `gurobi` of Gurobi 7.0.0 via the MEX file. This file provides an interface between MATLAB and `gurobi`, which is written in C++.

[15] This number of sub-problems is a result of 51 sub-problems (equivalent to the number of scenarios) times 29 L-shaped iterations.

[16] See Section 3.1.2 for a description of this heuristic.

considered in this work lead to about US$ 25 million of procurement costs. The different results are due to missing data in Rawls and Turnquist [2010]. For instance, distances between nodes, arc capacities $U^s_{(i,j)}$ and transport requirements u^k are not given. Moreover, information regarding demand is incomplete. According to Table 3 in Rawls and Turnquist [2010], demand is given for certain hurricanes but the corresponding landfall nodes are not specified. Such gaps had to be filled, leading to deviations in the data and hence also in solutions.

Figure 4.9 compares location decisions proposed by Rawls and Turnquist [2010] (top) and the suggested locations for the small-case case study based on the accelerated L-shaped method (bottom).[17]

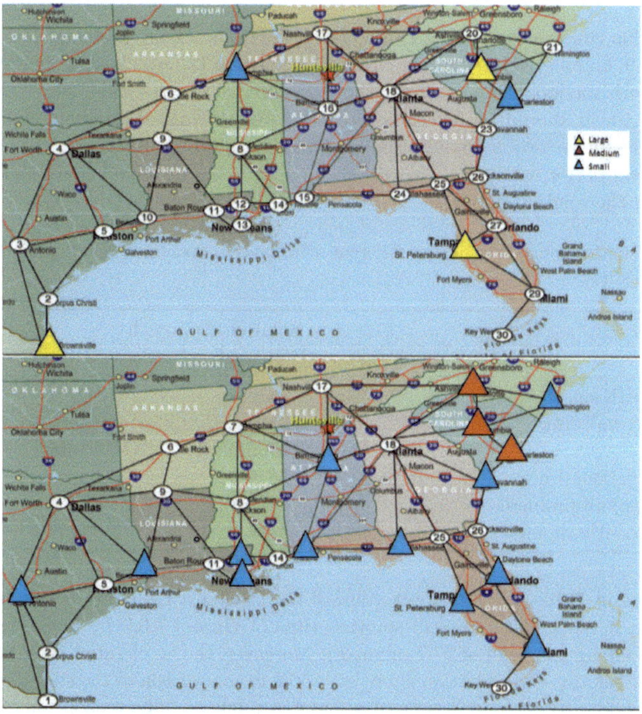

Figure 4.9: Facility locations as suggested by Rawls and Turnquist [2010] (top) and for the small-scale case study (bottom)

[17]Similar location results are obtained when the standard L-shaped method or Gurobi is used. The folder "Results" on https://zenodo.org/ contains the corresponding results for all three solution approaches [Grass, 2017].

In total, three large-sized and two small-sized facilities are opened in the case of Rawls and Turnquist [2010], incurring US$ 939,200 opening costs. In contrast, many more facilities of smaller size are located when the accelerated L-shaped method is used, resulting in US$ 820,000 fixed costs.[18] Some of these facilities are opened at vulnerable nodes (see Figure 4.3 for comparison). For instance, a facility is located in Charleston (node 22 in Figure 4.9 (top and bottom)) and in Miami (node 29 in Figure 4.9 (bottom)). However, several facilities located around these nodes can be used if these facilities are actually destroyed, e.g. the facility in Miami (node 29) can be replaced by the ones in Tampa (node 28) and Orlando (node 27).

Medium-Scale Case Study

The situation concerning solution times changes with increasing problem size. According to Table 4.9, Gurobi 7.0.0 solved the original problem in (2.1)-(2.9) after almost one hour, whereas both versions of the L-shaped method needed far less time (see "Overall time" in Table 4.9). It should be mentioned that a solution time of one hour for a problem consisting of almost two million decision variables is already satisfactory. The main reason for this good performance is the relatively small number of binary first-stage variables, simplifying the solution process.

Table 4.9: Performance overview of different solution approaches for the medium-scale case study

	Solution Approach		
	Gurobi	L-Shaped	Accelerated L-Shaped
MIPGap/RelOptGap	0.02	0.06	0.05
Overall time	3462 sec (57.7 min)	1348 sec (22.5 min)	631 sec (10.5 min)
Average time per sub-problem		0.7 sec	0.4 sec
No. iterations		16	16

As stated in Table 4.9, the application of the standard L-shaped method is already valuable as it reduced the overall solution time of Gurobi 7.0.0 by more than half an hour (22.5 minutes instead of 57.7 minutes). Moreover, the accelerated L-shaped method reduced the solution time additionally by over 50% in comparison to the standard L-shaped method, leading to an overall time of about ten minutes. The reason for the superior performance of the accelerated L-shaped method is that the specialized interior-point method SIPM solved each sub-problem almost twice as fast as MATLAB's interior-point method `linprog`, namely in 0.4 seconds instead of 0.7 seconds. Note that a similar relation

[18] As in the case of procurement costs, care must be taken when comparing the absolute values of opening costs due to missing data in Rawls and Turnquist [2010].

could already be observed for the small-scale case study, where SIPM was also nearly twice as fast as `linprog` (see Table 4.8).

In contrast to smaller problems, solving the master problem in each iteration of the L-shaped method does not represent the actual challenge for larger problems anymore. This can be seen in Figure 4.10, where `linprog` is more time-consuming, requiring a total time of 854.562 seconds for the sub-problems, than the function `gurobi` with about 435 seconds of solution time.[19] Therefore, solving the sub-problems efficiently is more crucial for larger problems, emphasizing the advantages of SIPM.

Profile Summary

Generated 28-Jun-2017 11:33:49 using performance time.

Function Name	Calls	**Total Time**	Self Time*
LShaped_MultiCut	1	1348.283 s	58.183 s
linprog	1280	854.562 s	1.397 s
optim\private\lipsol	1280	847.627 s	47.114 s
optim\private\lipsol>direction	60414	474.603 s	474.603 s
intlinprog_gurobi	16	434.962 s	0.044 s
gurobi (MEX-file)	16	434.918 s	434.918 s

Figure 4.10: MATLAB's profiler showing the most time-consuming functions of the standard L-shaped method for the medium-scale case study

Computational tests revealed the following effect: After the relative optimality gap RelOptGap in (4.1) fell below 0.05, it remained almost unchanged in the subsequent iterations. In this case the reduction of the gap between the lower LB and upper bound UB, defined in (4.2) and (4.3), respectively, is either weak or non-existent. Such a *tailing-off effect* is a well-known drawback of the L-shaped method [Desrosiers and Lübbecke, 2005]. As shown in Figure 4.11, the convergence of the lower and upper bound stagnates in later iterations of the L-shaped method, preventing RelOptGap in (4.1) from falling below the predefined accuracy level ϵ_O. Due to this stagnation both versions of the L-shaped method were terminated after 16 iterations with an achieved accuracy level of 0.06 and 0.05, respectively. For comparison reasons, Gurobi 7.0.0 was intended to terminate when MIPGap fell below 0.07. However, Gurobi 7.0.0 achieved a MIPGap of 0.02 in the end (see Table 4.9).[20]

[19] The complete profile summary can be found in the folder "Results" on https://zenodo.org/ [Grass, 2017].

[20] According to the log file "MediumLog" in Appendix A.6.2, MIPGap could be reduced from 0.183 to 0.0191 in the last iteration. The log file can also be found in the folder "Results" on https://zenodo.org/ [Grass, 2017].

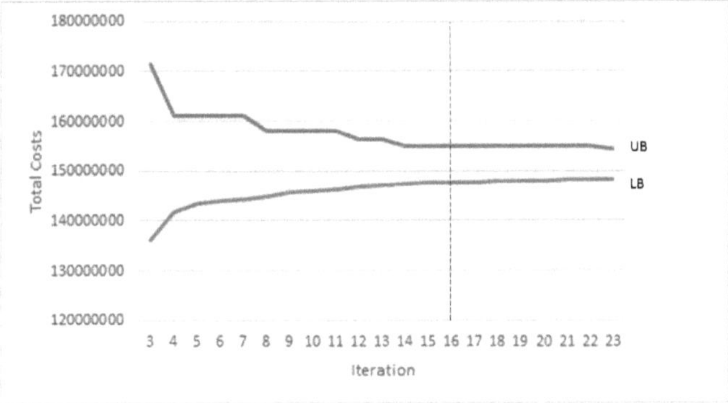

Figure 4.11: Upper bound (UB) and lower bound (LB) for the medium-scale case study

Although several techniques exist to overcome the tailing-off effect, most of these approaches are problem-dependent and have to be adjusted accordingly [Rahmaniani et al., 2017]. As the accelerated L-shaped method is intended to be generally applicable to a broad class of two-stage stochastic programs, no further adjustments have been made. Here, the main purpose is to show the advantages of the specialized interior-point method SIPM with respect to the reduced computation times of the sub-problems in comparison to the standard interior-point method `linprog`. Nevertheless, a multitude of possible improvements for the L-shaped method can be found in the literature [Rahmaniani et al., 2017] and can be implemented in addition to SIPM.

The location decisions for the medium-scale case study are postponed to the large-scale case study where location results for both case studies are compared and discussed.

Large-Scale Case Study

The previous case studies were aimed at illustrating the usefulness of the accelerated L-shaped method for an increased problem size. Moreover, it could be shown that SIPM outperformed `linprog` for both case studies. Now, the more realistic large-scale case study introduced in Section 4.1.1 is considered. Solving the Rawls and Turnquist [2010] model in (2.1)-(2.9) by Gurobi 7.0.0 required more than 64 hours, as shown in Table 4.10. Waiting almost three days for solutions is unacceptable, especially if fast reactions, e.g. in preparation for predicted disasters, are needed. It should also be mentioned that Gurobi 7.0.0 used 8 GB of memory, whereas the L-shaped method needed 2.8 GB on average for the large-scale case study. For relief organizations working with computers where available memory may be not sufficient for large-scale problems, the practical applicability of commercial solvers like Gurobi is therefore questionable.

Table 4.10: Performance overview of different solution approaches for the large-scale case study

	Solution Approach		
	Gurobi	L-Shaped	Accelerated L-Shaped
MIPGap/RelOptGap	0.004	0.009	0.008
Overall time	233,637 sec (64.90 h)	7541 sec (2.09 h)	3968 sec (1.10 h)
Average time per sub-problem		2 sec	0.98 sec
No. iterations		28	27

According to the "Overall time" in Table 4.10, using the standard L-shaped method led to a significant time reduction. In particular, a relative optimality gap RelOptGap of 0.009 could be achieved after just 2 hours. For comparison reasons, Gurobi was terminated as soon as MIPGap fell below 0.009 with a finally achieved optimality gap of 0.004.[21] When applying the accelerated L-shaped method the overall solution time could even be reduced by another hour and with a similar relative optimality gap of 0.008. In other words, the large-scale case study with almost five million decision variables could be solved in about one hour due to fast solutions of sub-problems via SIPM. According to the "Average time per sub-problem" in Table 4.10, SIPM needed about one second for each sub-problem on average. Just as in the case of the small- and medium-scale problem, solving the sub-problems with `linprog` took twice as long as SIPM (compare "Average time per sub-problem" in Table 4.10 with Table 4.8 and 4.9). Here, an average time of two seconds for the sub-problems could be obtained for `linprog`, which already seems to be fast. However, 118 sub-problems (equivalent to the number of scenarios) have to be solved in each iteration of the L-shaped method (with 28 iterations in total), resulting in an overall number of 3,304 sub-problems. As shown in Figure 4.12, about two hours (6,673.440 seconds) were needed for MATLAB's built-in function `linprog` to solve these sub-problems. Remarkably, the solution of 28 master problems consumed only about ten minutes (about 609 seconds for `intlinprog_gurobi` or `gurobi` in Figure 4.12).

[21] Similar to the medium-scale case study, MIPGap jumped from 0.173 to 0.0043 in the last iteration (see Gurobi's log file "LargeLog" in Appendix A.6.3 or the folder "Results" on https://zenodo.org/ [Grass, 2017]).

Profile Summary
Generated 22-Jul-2017 15:23:23 using performance time.

Function Name	Calls	Total Time	Self Time*
LShaped_MultiCut	1	7540.951 s	258.061 s
linprog	3304	6673.440 s	5.347 s
optim\private\lipsol	3304	6646.820 s	320.435 s
optim\private\lipsol>direction	174752	4370.933 s	4370.933 s
optim\private\lipsol>getpu	90680	1029.906 s	727.013 s
spdiags	178056	617.749 s	617.749 s
intlinprog_gurobi	28	608.871 s	0.555 s
gurobi (MEX-file)	28	608.317 s	608.317 s

Figure 4.12: MATLAB's profiler showing the most time-consuming operations of the standard L-shaped method for the large-scale case study

The same computing time for the master problems can be confirmed in the case of the accelerated L-shaped method. As shown in Figure 4.13, ten minutes (about 605 seconds) were also required to solve the master problems within the accelerated version. In contrast, the time consumption of SIPM, namely 3,107.240 seconds, accounted for the major part of overall solution times. The comparison of the "Total Time" of SIPM in Figure 4.13 and `linprog` in Figure 4.12 reveals the superiority of SIPM concerning computing times for the sub-problems.

Profile Summary
Generated 21-Jul-2017 17:27:32 using performance time.

Function Name	Calls	Total Time	Self Time*
LShaped_MultiCut	1	3967.562 s	247.272 s
SIPM	3186	3107.240 s	2706.724 s
intlinprog_gurobi	27	605.428 s	0.552 s
gurobi (MEX-file)	27	604.876 s	604.876 s
spdiags	238986	380.959 s	380.959 s

Figure 4.13: MATLAB's profiler showing the most time-consuming operations of the accelerated L-shaped method for the large-scale case study

As emphasized in Section 3.3.3, determining the Newton search direction in each iteration of an interior-point method is computationally most expensive. This can also be seen in

Figure 4.12, where the function **direction**[22] was most time-consuming. In contrast, the most time-consuming steps of SIPM are shown in Figure 4.14.[23] Due to the relatively small size of the Schur complement $\tilde{\mathcal{S}} \in \mathbb{R}^{450 \times 450}$ in (3.47) the Cholesky factorization **chol** was the third most expensive operation. A comparable amount of time was spent for the computation of the (2,2)-block matrix B in (3.44), whereas determining the Schur complement $\tilde{\mathcal{S}} = B - CA^{-1}C^T$ was computationally most expensive. All of these operations had to be executed in each L-shaped iteration and for each scenario, resulting in 119,493 "Calls" in Figure 4.14. In Section 4.3, approaches for dealing with these time-consuming steps are given for the case that the Schur complement $\tilde{\mathcal{S}}$ in (3.47) becomes very large for some applications.

Lines where the most time was spent

Line Number	Code	Calls	Total Time	% Time	Time Plot
122	`Schur=B-(C*AC); % Schur comple...`	119493	928.725 s	29.9%	
106	`B=HQ*(spdiags(zg,0,n,n)*HQ');`	119493	662.174 s	21.3%	
123	`UT=chol(Schur); % Using the Ch...`	119493	622.486 s	20.0%	
125	`d_pi_2=UT\(UT'\(rhs_2-C*(Amat\...`	119493	137.156 s	4.4%	
101	`Du_diag=sparse(ii',jj(:),Du_re...`	119493	132.327 s	4.3%	
All other lines			624.372 s	20.1%	
Totals			3107.240 s	100%	

Figure 4.14: MATLAB's profiler showing the most time-consuming operations of SIPM for the large-scale case study

A common drawback of interior-point methods occurs at the end of the solution process when $\mu \longrightarrow 0$, leading to an ill-conditioned system, i.e. to a large condition number (see page 69). Figure 4.15 shows the condition number[24] of the saddle point matrix S in (3.41) for each iteration of the interior-point method for one scenario of the large-scale case study. Moreover, the corresponding relative primal and dual residuals as well as the relative optimality gap, all defined in (3.39), are given.

[22]The function `optim\private\lipsol>direction` is a child function of MATLAB's interior-point method `lipsol` (`optim\private\lipsol`), which is in turn the child function of `linprog`.

[23]The complete profile summaries for Figure 4.12, 4.13 and 4.14 can be found in the folder "Results" on https://zenodo.org/ [Grass, 2017].

[24]MATLAB's built-in function `condest` was used to determine the condition number of the sparse saddle point matrix S in (3.41) and multiplied by `eps` for comparison reasons in Figure 4.15. The function `eps` represents the floating-point relative accuracy, measuring the distance between 1 and the next larger double-precision number, i.e. `eps`=2.2204e-16.

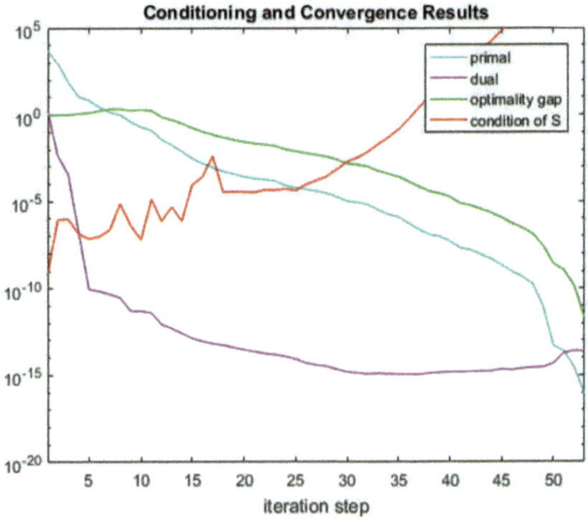

Figure 4.15: Condition number of saddle point matrix S, relative residuals and optimality gap

The first observation is that the relative dual residual (pink line) converges much faster to the desired accuracy level than the relative primal residual (blue line) and the relative optimality gap (green line). Second, when the interior-point method approaches the solution, i.e. a higher level of accuracy is reached, the condition number (red line) increases significantly, leading to a severely ill-conditioned saddle point matrix S. Developing an efficient solution method for such ill-conditioned problems is a challenging task [Benzi et al., 2005] and an important future research direction, as will be outlined in Section 4.3.

In general, calculation times of the sub-problems could be decreased even further if a more sophisticated programming language like Fortran, as in the case of `linprog`, or C++ instead of MATLAB was used, and if sub-problems were solved in parallel.

Location Decisions for the Large-Scale Case Study

In the following, location results for the more realistic case study are presented and compared with the heuristically found solution of Rawls and Turnquist [2010]. Since most of their parameter values differ from the ones defined in the newly proposed case study, the resulting solutions and overall costs are difficult to compare. Nevertheless, practical implications of location decisions for both case studies can be demonstrated, as fixed opening costs F_l, capacities M_l and the general cost structure are equal in both case studies. In particular, transportation costs $c^k_{(i,j)}$ increase linearly with increasing distance and the penalty costs h^k and p^k have the same ratio as in Rawls and Turnquist [2010] (see Section 4.1.1, page 76).

As already mentioned on page 88, almost US$ 1 million total opening costs arise in the case of Rawls and Turnquist [2010]. In contrast, location solutions for the large-scale case study stipulate many more small-sized facilities, as will be shown below, resulting in opening costs of US$ 431,200. Due to the high costs for large-sized facilities it is cheaper to open more smaller facilities, as also the small case study showed.[25] In contrast, costs for pre-positioning are much higher in the case of the large-scale case study, namely US$ 55.6 million in comparison to US$ 29.5 million in Rawls and Turnquist [2010]. The reasons are two-fold. First, more types of relief items have to be purchased with cost rates different from Rawls and Turnquist [2010]. Second, more overall demand has to be satisfied since more areas are defined as potentially affected in the newly proposed case study. As a further result, transportation costs as well as penalty costs for unused items and unsatisfied demand are also much higher.

In order to illustrate the consequences if more historical hurricanes are used for the scenario definitions, location decisions given in Rawls and Turnquist [2010] (see Figure 4.9 (top)) are included in Figure 4.1, where differently affected counties based on hurricanes since 1900 are shown. The result is presented in the top of Figure 4.16. In addition, locations for the medium-scale case study (center) and the large-scale case study (bottom) are shown in Figure 4.16 based on the results of the accelerated L-shaped method.

In the case of Rawls and Turnquist [2010], the two southern large-sized facilities in Figure 4.16 (top) are opened in Brownsville and Tampa, which are under particular threat, indicated by the dark violet color. These regions are not classified as potentially vulnerable in Rawls and Turnquist [2010] and Figure 4.3, but they are when taking all hurricanes of the last 115 years into account. One can deduce that these two southern facilities, and the relief items stored there, are likely to be destroyed in a future hurricane setting. In this case, items have to be supplied from the facility located in Memphis (blue triangle in the north). However, some serious injuries have to be treated with medicine within the first four hours, which cannot be guaranteed in this case. Even if some items are not time-critical, sufficient quantities may not be available due to the small size of this facility. In contrast, more widely spread small-sized facilities are opened in the case of the medium- and large-scale problem. Note that the medium-scale case study is "incomplete", i.e. only 80 of 118 originally defined scenarios are used. For this reason, some facilities are located in vulnerable areas (dark violet), e.g. the blue triangle in the south of Florida. However, location decisions for the medium-scale problem already represent a part of the final solution for the large-scale case study.[26] As shown in Figure 4.16 (center), many more facilities than in the case of Rawls and Turnquist [2010] are opened and located along the coast. This medium-scale solution is expanded for the large-scale case study by locating additional small-sized facilities along the coast and the northern border. The latter can be used as backup storages if coastal facilities are destroyed. According to Figure 4.16

[25] In the small-scale case study the opening costs amounted to US$ 820,000, i.e. almost twice the costs of the case study considered here. This is mainly due to the three medium-sized facilities (see Figure 4.9 (bottom)).

[26] For the medium-scale case study the opening costs are US$ 147,400 and pre-positioning costs US$ 24,201,480.30.

(bottom), relief facilities are generally located away from vulnerable areas, but still close enough to provide immediate assistance. For instance, the most southern facility (near Miami in Figure 4.16 (center)) of the medium-scale problem is not part of the location set of the large-scale case study anymore. Another example is given by the facility at the most western point (near Corpus Christi in Figure 4.16 (bottom)), which is located near to the vulnerable region but in contrast to the case of Rawls and Turnquist [2010] it is sufficiently far away. However, the second facility in the west (near Houston in Figure 4.16 (bottom)) is located directly in the vulnerable area. Even if both facilities in the west are destroyed by a hurricane, additional facilities can be found nearby. Due to the widespread availability of storage facilities for the newly proposed case study, distances are shortened, risk of destruction is reduced and coverage of affected people can be improved.

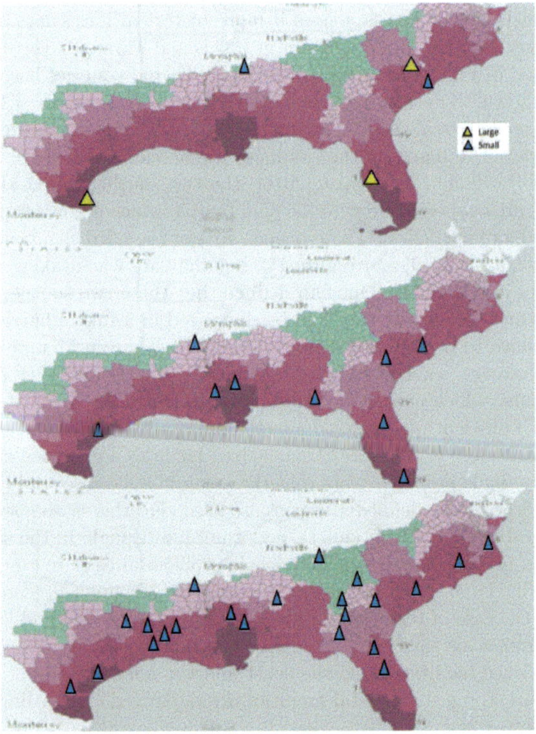

Figure 4.16: Facility locations as suggested by Rawls and Turnquist [2010] (top) and for the medium-scale case study (center) and the large-scale case study (bottom)

The extremely different location solutions of Rawls and Turnquist [2010] and the large-scale case study are mainly due to the way in which scenarios are defined, namely which areas are identified as particularly vulnerable. In Figure 4.1 the region under consideration is divided into more differentiated and partially different vulnerability classes than in the case of Rawls and Turnquist [2010] in Figure 4.3. As a result, scenario definitions and therefore the location solutions differ between both case studies. Moreover, if scenarios do not cover the full range of possible outcomes, as in the case of the medium-scale case study, facilities may be located at potentially vulnerable nodes. In summary, scenario definitions not sufficiently reflecting reality can have far-reaching implications with regard to the performance of humanitarian operations. Therefore, designing realistic scenarios and case studies may have life-saving consequences. The downside is that the resulting problems are generally very large and are harder to solve even with today's commercial solvers.

In this section it was shown that exploiting the two-stage structure via the L-shaped method and making use of the special structure of the recourse matrix W in (3.5) via SIPM leads to significant time savings for large problem sizes. Although strategic location and pre-positioning decisions may not be time-critical, a commercial solver like Gurobi may even fail to provide feasible solutions for realistic large-scale problems. For such problem sizes the L-shaped method is therefore a reasonable and necessary choice. Accelerating the L-shaped method is of particular importance when relief organizations want to initiate preparation activities for disasters predicted a few days in advance like in the case of hurricanes. One of the most destructive hurricanes in the history of the United States was hurricane Katrina in August 2005. Temporarily, Katrina reached a wind speed of 280 km/h and caused over 1800 casualties [Knabb et al., 2011]. Louisiana, and especially New Orleans, were among the most severely affected regions, suffering great damage due to two dike bursts. Although the *National Hurricane Center* NHC issued a warning that Katrina would become a major hurricane and would hit mainland in three days, one of the most renowned relief agencies, the *American Red Cross*, failed to pre-position a sufficient amount of relief items [Elliott et al., 2014]. In order to show the advantage of using the Rawls and Turnquist [2010] model in such a situation and solving it by the accelerated L-shaped method, a case study is designed in the next section based on the forecast provided by NHC three days before Katrina's landfall. Results concerning locations and computation times are presented in Section 4.2.2.

4.2 Case Study Based on a Hurricane Forecast

The purpose of this section is as follows. First, relief organizations often store aid supplies at temporary facilities only when a specific hurricane is predicted. Hence, location and pre-positioning decisions provided by the Rawls and Turnquist [2010] model are evaluated based on the actual path and destruction level of Katrina, emphasizing the advantage of using this two-stage stochastic model in the short-term preparation phase. Second, this case study is used to analyze how solution times change if one part of the input data is changed. In particular, this case study has the same problem size as the large-scale case study, i.e. $|K| = 5$, $|L| = 3$, $|N| = 90$ and $|S| = 118$ (see Table 4.7 in Section 4.1.1). In addition, the scenario-independent parameters like cost rates and capacity requirements are taken from Section 4.1.1. The only difference between the large-scale case study and the one proposed here is the definition of the scenario-dependent parameters P_s, $\rho_i^{k,s}$, $v_i^{k,s}$ and $U^s_{(i,j)}$, which are tailored to the forecast for hurricane Katrina as described next.

NHC provides forecasts for all tropical cyclones in the Atlantic and Eastern Pacific basin. In particular, hurricane forecasts are given 3-5 days before a hurricane's landfall, with wind speeds and potential paths being updated every six hours. In very critical cases even more frequent updates are available. Relief organizations receive the relevant information from the *Hurricane Liaison Team* (HLT) or the local *Weather Forecast Offices* (WFO). "*The HLT is a Federal Emergency Management Agency (FEMA)-sponsored team made up of federal, state, and local emergency managers, FEMA personnel, and specially trained National Weather Service (NWS) forecasters. They work to serve as a 'bridge' between NHC forecasters and local emergency managers dealing with a tropical cyclone threat*" [NHCFP, 2017]. Based on the given information, emergency managers can initiate preparation activities like the pre-positioning of essential aid supplies.

4.2.1 Data

In Figure 4.17, the 3-day cone of hurricane Katrina is shown, illustrating the current position of the hurricane center on August 26, 2005 at 5pm and the potential path of this center for the next three days.

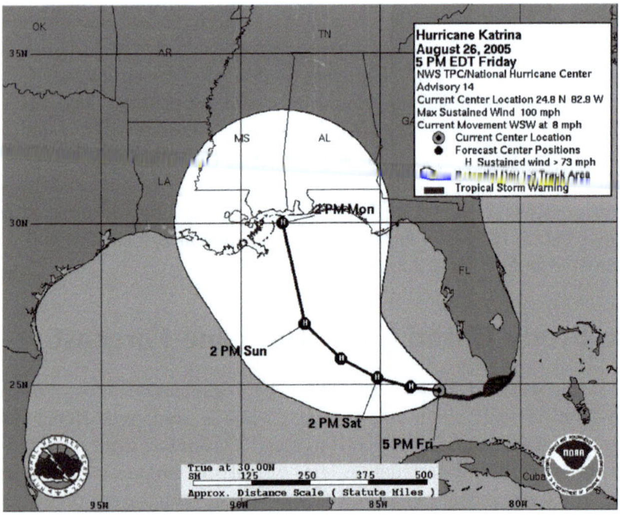

Figure 4.17: Predicted path of hurricane Katrina, 26 August 2005 at 5pm (three days before landfall [NHC, 2005])

The cone describes the range in which the hurricane center could move. As one can see, mainly four states Louisiana (LA), Mississippi (MS), Alabama (AL) and Florida (FL) are under particular threat. As stated in Section 4.1.1, the hurricane center has a strike circle of 80.5 km radius on average. So if the actual center would move to the border of the predicted cone, areas within a 80.5 km radius around the hurricane center can be identified as being at risk. For the case study considered here, nodes within the predicted cone plus 80.5 km around it are assumed to be vulnerable, as shown in Figure 4.18.

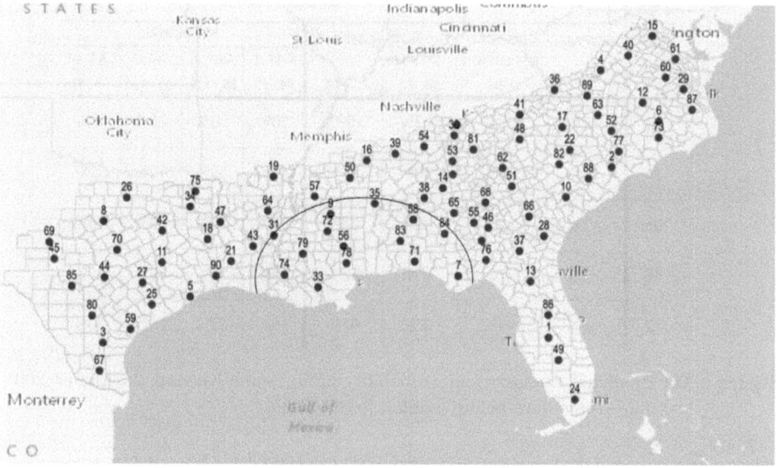

Figure 4.18: Potentially vulnerable nodes based on the predicted path of hurricane Katrina

The network is the same as in the large-scale case study, where (as before) facilities of three sizes can be established. However, it is assumed that, unlike in the previous case study, demand can only arise at nodes within the vulnerable area, labeled in Figure 4.18. The definition of demand depends on the respective scenario, containing the information on Katrina's potential location and damage level. In total, 118 scenarios are considered. In contrast to the 118 scenarios of the large-scale case study, these scenarios are defined in a much more differentiated manner. For instance, Katrina can hit node 33 in Figure 4.18 with a specific intensity and then be diverted westwards affecting node 74 or eastwards striking nodes 78 and 56. As wind speed can change significantly within a few hours these nodes can be hit by different categories, leading to a variety of more detailed scenarios.

Fortunately, and in contrast to the case study designed previously, probability of occurrence is provided by NHC directly. In Figure 4.19 forecasts concerning wind speeds are given for the next 72 hours. For instance, NHC predicted that Katrina will become a category 4-5 hurricane with a probability of 30% in 72 hours.[27]

WIND SPEED FORECAST FOR KATRINA
EXPRESSED AS PROBABILITY
FROM NHC ADVISORY 14
5:00 PM EDT AUG 26 2005

TIME	WIND SPEED INTERVAL IN MPH							
HOURS	DISSIPATED	TROPICAL DEPRESSION < 39	TROPICAL STORM 39 - 73	HURRICANE >= 74	HURRICANE CAT. 1 74 - 95	CAT. 2 96 - 110	CAT. 3 111 - 130	CAT. 4-5 >= 131
12	<2%	<2%	<2%	>98%	30%	50%	20%	3%
24	<2%	<2%	5%	95%	40%	30%	15%	5%
36	<2%	<2%	<2%	>98%	30%	25%	30%	15%
48	<2%	2%	2%	95%	35%	15%	20%	20%
72	<2%	<2%	2%	97%	15%	25%	25%	30%

Figure 4.19: Predicted category and probability of hurricane Katrina, 26 August 2005 at 5pm (three days before landfall [NHC, 2005])

For each scenario, the scenario-dependent parameters $\rho_i^{k,s}$, $v_i^{k,s}$ and $U_{(i,j)}^s$ are specified as described in Section 4.1.1. For instance, $\rho_i^{k,s}$ is set to 0.5 if node i is assumed to be hit by a category 3 hurricane and 0 for a category 4 or 5 hurricane. As it is highly unlikely that Katrina will destroy facilities located outside the vulnerable area in Figure 4.18, $\rho_i^{k,s}$ is set to 1 for these nodes. Similarly, the arc capacity $U_{(i,j)}^s$ is only reduced to 0 if the respective arc (i,j) is within the predicted hurricane region and is hit by a hurricane 4 or 5 category.

As time is very limited in the preparation phase for a specific hurricane, data input should be made very quickly. Scenario-independent parameters can be determined even before the hurricane season starts and can be adjusted only if necessary. In contrast, parameters depending on scenarios have to be specified for each particular hurricane forecast. In order to help aid organizations with this time-consuming task, the Excel

[27] Each probability for a specific hurricane category is calculated based on historical observations, Monte-Carlo simulations and forecast errors [NHC, 2014]. The final probabilities for each hurricane category in Figure 4.19 are rounded to the nearest 5% (except less than 5%) [Landsea, 2017]. Therefore, the sum of probabilities for each hurricane category (e.g. 95% in the last row) does not exactly comply with the total probability that a hurricane will occur in the next 72 hours, i.e. 97%.

file *KatrinaCaseStudy.xlsm* can be used.[28] Here, endangered nodes, hurricane category and corresponding probabilities of occurrence have to be defined according to the NHC forecasts. Afterwards, scenario-dependent parameters $\rho_i^{k,s}$, $v_i^{k,s}$ and $U_{(i,j)}^s$ are updated automatically.

4.2.2 Computational Results

Table 4.11 gives the performance overview of all three solution approaches for the case study based on the three-day forecast. After almost 30 hours a solution with a relative optimality gap of 0.003 could be found by Gurobi 7.0.0. Bearing in mind that relief items have to be ordered, picked, loaded and delivered to the respective destination, aid agencies cannot afford to wait more than one day for a solution. In contrast, exploiting the two-stage structure via the standard L-shaped method led to an overall solution time of about one hour, highlighting the usefulness of this method for large-scale problems. However, the accelerated version was even able to reduce the overall solution time to 38 minutes. Obtaining solutions in such a short time paves the way for relief organizations to use two-stage stochastic models in preparation for acute threats.

Table 4.11: Performance overview of different solution approaches for the case study based on the short-term forecast

	Solution Approach		
	Gurobi	L-Shaped	Accelerated L-Shaped
MIPGap/RelOptGap	0.003	0.005	0.006
Overall time	105,651 sec (29.35 h)	4,114 sec (1.14 h)	2,281 sec (38 min)
Average time per sub-problem		1.8 sec	0.9 sec
No. iterations		18	18

According to Table 4.11, overall solution times were shorter than for the large-scale case study despite the same problem size (see "Overall Time" in Table 4.10 for comparison). This can mainly be explained by the fact that only a relatively small portion of nodes and arcs of the original network is actually relevant for the solution. As already described, most nodes in Figure 4.18 lie outside the potentially vulnerable area and therefore no demand and damage arise at these nodes. It is of course possible to define a new network consisting of 90 nodes but which covers solely counties within the hurricane cone and its neighborhood (as given in Figure 4.17 and 4.18). Note that in this case similar solution times as for the previously designed case study can be expected, i.e. almost 65 hours if Gurobi is used and only about one hour for the accelerated L-shaped method. However, the main purpose of

[28] The Excel file *KatrinaCaseStudy.xlsm* can be found in the folder "DataImport" on https://zenodo.org/ [Grass, 2017].

this subsection is to analyze the computational performance of the accelerated L-shaped method, and in particular of SIPM, if the basic case structure remains the same but if only scenario definitions with corresponding scenario-dependent parameters are changed.

The success of the accelerated L-shaped algorithm is due to the use of SIPM for the sub-problems. As indicated by "Average time per sub-problem" in Table 4.11, exploiting the specific structure of W in (3.5) led to an average time of 0.9 seconds, which is again half of the time needed by `linprog`. As for the previous case studies, i.e. independent of the problem size and data, SIPM was consistently twice as fast as `linprog` (see Table 4.8, 4.9 and 4.10 for comparison).

Location Decisions for the Hurricane Katrina Case Study

Solving the hurricane Katrina case study via the accelerated L-shaped method led to the locations shown in Figure 4.20.

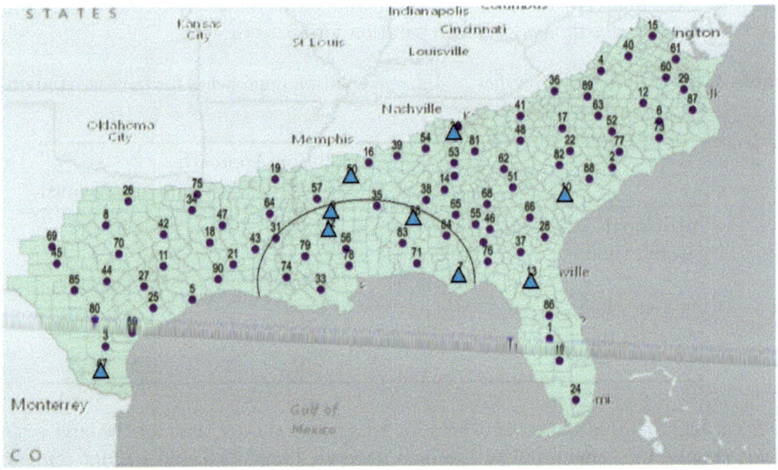

Figure 4.20: Suggested facility locations for hurricane Katrina three days before landfall

As can be seen in this figure, nine small-sized facilities are opened within or in proximity to the hazardous area predicted by NHC (see Figure 4.17 and 4.18). If Katrina were to make landfall more in the East, destroying facilities at node 7 and/or 58, facilities in the East (nodes 10 and 13), West (nodes 72 and 9) or North (nodes 23 and 50) could provide first-aid supplies. In the contrary case, if the westernmost coast within the predicted hurricane cone is affected (nodes 9 and/or 72), relief items can be transported from facilities located at nodes 50, 58 and 7. Note that two facilities, one in the North (node 30) and one in the West (node 67), lie far away from the predicted area. Numerical tests showed that these facilities no longer belonged to the solution set if a lower RelOptGap than 0.006 is achieved.[29] For a more accurate solution, e.g. where RelOptGap is 0.003, a facility is chosen at node 34 (instead of 67) and at node 38 (instead of 30).

Ex-Post Analysis

On August 29, 2005 Katrina hit the coast of Louisiana near the Mississippi border as a category 3 hurricane with a wind speed of 195 km/h [Knabb et al., 2011]. Counties affected by Katrina and suggested facility locations are illustrated in Figure 4.21.

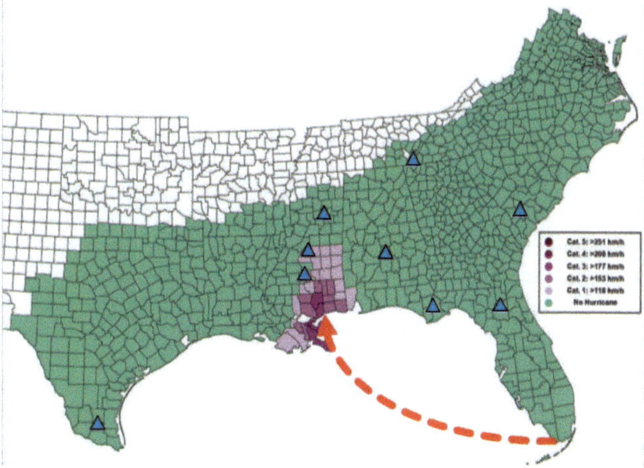

Figure 4.21: Actual path of hurricane Katrina (red dashed line), counties affected to varying extents and suggested facility locations

Although the intensity of hurricane Katrina weakened to the lowest category while moving inland, its effects were devastating. The main cause of the high level of destruction and the high number of fatalities was the immense storm surge produced by this hurricane. The high-water mark of almost 9 meters was reached at the coast, flooding large parts of New Orleans (node 33 in Figure 4.20). In particular, 80% of the city was up to 6 meters under water [Knabb et al., 2011]. Moreover, heavy rainfall of up to 25-30 cm in Louisiana and Mississippi and 20 cm more inland increased the flood level and exacerbated the already difficult and chaotic situation. Flooding caused the most fatalities and destroyed homes, buildings and entire infrastructures. Roads, highways and some bridges in the coastal area were damaged, cutting off especially the area around New Orleans.

[29] Due to the tailing-off effect, as discussed on page 89, both versions of the L-shaped method were terminated after 18 iterations.

In spite of a relatively accurate forecast for hurricane Katrina, the American Red Cross was not able to meet the basic needs of those affected. For instance, neither water nor food could be provided by the relief organization in, e.g., Mississippi [The New York Times, 2006]. In contrast, the suggested locations in Figure 4.21 could have provided aid supplies within a few hours from the nearest facilities. One can assume that facilities at nodes 9 and 72 were not destroyed by the category 1 hurricane (areas in light pink). However, access to these facilities were likely to be limited by heavy rain and flooding. In this case relief items could be transported from facilities located at nodes 50 and 58, still guaranteeing fast supply.

Even nowadays, relief organizations use hurricane forecasts to locate facilities in a rather intuitive manner, simply near the predicted hurricane cone. They argue that the effort of implementing and solving an optimization model is higher than its benefits [ISCRAM, 2017]. However, naive approaches of relief agencies cannot capture the entire complexity of the pre-positioning decision problem, leading to inadequate preparation measures as in the case of the American Red Cross. In contrast, the two-stage stochastic model proposed by Rawls and Turnquist [2010] combined with historical hurricane data finds the cost optimal solution, taking all possible disaster outcomes with the corresponding probability of occurrence into account. Location and pre-positioning decisions are made considering potential destruction of arcs and facilities. The advantage of this model is the simultaneous minimization of different costs arising before and after the occurrence of a disaster. Preparation decisions based solely on experience or rules of thumb cannot guarantee such a cost efficient and at the same time demand satisfactory solution.

When aid agencies act on three-day forecasts and would like to use a two-stage stochastic model, fast solution times are indispensable. In this section it was shown that the accelerated L-shaped method provides location and pre-positioning solutions after about half an hour. Obtaining solutions within such a time frame enables aid agencies to react in a more flexible and agile manner to disaster forecasts available in the short-term.

4.3 Outlook

The previous computational results showed the advantage of exploiting the specific structure of the recourse matrix W in (3.5) with respect to solution times. The specialized interior-point method SIPM could cut down the solution times for the sub-problems by about half of the time needed by `linprog` (see Tables 4.8 - 4.11). As described in Section 3.3.2, the accelerated L-shaped method can be applied to a broad class of two-stage stochastic programs. However, the Schur complement $\tilde{S}$ in (3.47) can be very large for some problem formulations. As highlighted in Figure 4.14, formulating and factorizing $\tilde{S}$ is most time-consuming and may become too expensive for a very large $\tilde{S}$. In such cases, it may be more reasonable to solve the saddle point system (3.41) instead of (3.47) to obtain the search direction in every iteration of the interior-point method. Benzi et al. [2005] provide a thorough description of properties and solution algorithms of saddle point problems arising from different application areas. In the case of interior-point methods, solving the saddle point system (3.41) is affected by the more and more ill-conditioned diagonal matrix D within the saddle point matrix S. As already described, the diagonal elements of the (1,1)-block matrix D tend to 0 or infinity when the solution is approached, increasing the condition number[30] of S.

Especially for large and sparse linear systems like (3.41), iterative procedures like Krylov subspace methods[31] are often preferred over direct approaches [Keller et al., 2000]. This is because storage requirements and computational costs of exact methods may be too expensive for large-scale problems. For instance, the Cholesky factorization requires $n^3/3$ flops to decompose a matrix of size $n \times n$ for most algorithms [Quarteroni et al., 2006, p. 83].[32] In other words, the computational effort increases cubically with an increasing problem size. However, using iterative algorithms like Krylov subspace methods to solve (3.41) at the end of an interior-point method, i.e. when S becomes ill-conditioned, requires a suitable preconditioner. In general, a preconditioner $P \in \mathbb{R}^{n \times n}$ is used to approximate the matrix $A \in \mathbb{R}^{n \times n}$ such that solving the preconditioned system

$$P^{-1}Ax = P^{-1}b,$$

with solution vector $x \in \mathbb{R}^n$ and right hand side $b \in \mathbb{R}^n$, is less expensive than solving the original system $Ax = b$ [Golub and Van Loan, 2013, p. 598]. Appropriate preconditioning generally reduces the condition number, leading to faster convergence of iterative methods. There exists a wider range of preconditioners for saddle point systems (3.41) than for the normal equations (3.43) [Al-Jeiroudi et al., 2008] as the condition number is squared in the latter case [Mathew, 2008, p. 448]. Note that preconditioners proposed for ill-conditioned systems (3.41) and (3.43) are usually suitable for final iterations of interior-point methods and perform poorly in early iterations [Bocanegra et al., 2013]. Therefore, one may switch preconditioners within interior-point methods according to the condition number of the respective matrix. As stated by Bocanegra et al. [2007], most preconditioners proposed for saddle point problems arising in interior-point methods are more appropriate for non-linear and quadratic than for linear programs. This highlights the need for developing preconditioners which are suitable for linear programming problems, e.g. arising in humanitarian contexts. In the case of two-stage stochastic programs in disaster management, the use of Krylov subspace methods with efficient preconditioners could possibly allow the solving of even larger sub-problems in an acceptable amount of time.

Besides developing the specialized interior-point algorithm further, the L-shaped method can be improved. In the classical version of the L-shaped method (Figure 3.2) the (mixed) integer master problem is solved to optimality in each L-shaped iteration after an optimality cut (3.17) is added. These computationally expensive steps, especially for large-scale master problems, can be reduced by applying the so-called *one-tree approach* [Rahmaniani et al., 2017]. The main idea is to construct a single branch-and-bound tree where each node

[30]See page 69 for a definition of the condition number.

[31]A thorough introduction to Kylov subspace methods can be found in Saad [2003].

[32]One flop represents one floating-point operation like plus, minus, divide and multiply [Quarteroni et al., 2006, p. 53].

is visited only once [Naoum-Sawaya and Elhedhli, 2013]. At each node the LP relaxation[33] of the master problem is solved and the corresponding first-stage solutions are used to solve the sub-problems for generating additional optimality cuts. Therefore, computing the integer solution of the master problem in every L-shaped iteration can be avoided. A similar procedure is used by the lazy constraint callback of CPLEX described on page 55.

Another interesting approach for improving the L-shaped method is to generate stronger optimality cuts, accelerating the overall solution process. According to Rahmaniani et al. [2017], several techniques exist in the relevant literature which can construct tighter cuts reducing the total number of L-shaped iterations. However, such approaches depend highly on the underlying problem formulation. Therefore, developing enhanced cut generation procedures for a broad class of two-stage stochastic models in the humanitarian context represents an important future research direction.

[33] A (mixed) integer linear program is called LP relaxation if the integrality constraints are omitted [Vanderbei, 2014, p. 352]. Binary constraints are relaxed by allowing the variables to take values between 0 and 1.

5 Conclusion

Countries all over the world, independent of their economic and financial prosperity, are suffering severely from the consequences of natural disasters every year. Floods and storms occur most frequently, causing damage running into billions and numerous deaths. Although natural disasters are not preventable, floods and storms are at least predictable. For instance, hurricanes can be forecasted three to five days in advance, allowing aid agencies to initiate preparation measures. Such preparation decisions have still to be made under uncertainty, since the actual path of the hurricane, its landfall point and category may deviate from the forecast. Nevertheless, even rough estimations are sufficient to set up shelters for evacuees and to pre-position relief items in existing public buildings like gyms and town halls. Often aid agencies use rules of thumb and experience to decide where to store relief items in preparation for a particular threat. Such intuitive decisions can have fatal consequences regarding insufficient inventory levels, as discussed in Section 4.2.2 in the case of the American Red Cross. Despite numerous accomplished relief operations in the past, the American Red Cross could not ensure the adequate and timely supply of urgently needed goods after Katrina's landfall in 2005. Decisions taken in an emergency situation are highly complex, involving various constraints and a high degree of uncertainty. Two-stage stochastic programs, often proposed in the humanitarian literature, can significantly support decision makers in this regard. Such optimization models are only useful in practice if corresponding solutions can be quickly obtained. The state-of-the-art commercial solver Gurobi 7.0.0 needed almost three days for the solution of the large-scale case study (see Table 4.10 in Section 4.1.3) and more than one day to solve the case study based on the three-day forecast for hurricane Katrina (see Table 4.11 in Section 4.2.2). Losing one to three days for calculations is unacceptable since relief items have to pass an entire supply chain before they are available at the respective destination. The accelerated version of the standard L-shaped method could reduce the running time to about one hour for the large-scale case study and half an hour for Katrina's case study. Such short solution times allow the use of two-stage stochastic models even in preparation for acute threats.

In order to provide acceptable solutions within a reasonable time frame, many researchers develop heuristics for the proposed two-stage models. However, heuristics have many disadvantages which were identified in Section 3.1.2. Besides the general inability to measure the quality of the heuristically found solution, a heuristic is highly problem-dependent and requires background knowledge of the problem for its sensible application. It is very unlikely that relief organizations have such expertise, preventing them from applying two-stage stochastic models in practice. In contrast, no parameter adjustments and background knowledge about mathematical optimization are needed for the accelerated L-shaped method. Moreover, the algorithm can be terminated whenever the solution

E. Graß, *An Accelerated Solution Method for Two-Stage Stochastic Models in Disaster Management*, Mathematische Optimierung und Wirtschaftsmathematik | Mathematical Optimization and Economathematics,
https://doi.org/10.1007/978-3-658-24081-3_5

achieves a desired optimality level, i.e. the solution quality is known. Finally, the newly proposed solution method can be used for a whole class of two-stage stochastic programs. As stated in Section 3.3.2, most of the models reviewed in this thesis meet the criteria for applying the accelerated L-shaped method. In summary, this newly proposed solution algorithm overcomes the drawbacks of heuristics and provides fast and high-quality solutions, for a whole class of two-stage stochastic models.

The success of the accelerated L-shaped algorithm is due to its key ingredient, the specialized interior-point method SIPM introduced in Section 3.3.3. Numerical tests revealed that SIPM required half the solution time of the standard interior-point method `linprog` for the sub-problems. Regardless of the problem size and data, SIPM outperformed `linprog` consistently (see Tables 4.8, 4.9, 4.10 and 4.11). The main idea of SIPM is to exploit the special structure of the recourse matrix W in (3.5). To be be even more specific, the most time-consuming step within each interior-point method, the determination of Newton's search direction, can be significantly simplified by reducing the problem size. Instead of solving the saddle point system in (3.41) or the system of normal equations in (3.43), the search direction is determined by the much smaller system in (3.47). Due to the special block triangular structure of W in (3.5), the dimension of the saddle point matrix $\tilde{\mathcal{S}}$ in (3.47) could be considerably reduced, allowing the use of the Cholesky factorization.

For problems where $\tilde{\mathcal{S}}$ in (3.47) is very large alternatives to direct methods like Cholesky have to be found. Section 4.3 gave an outlook of how iterative algorithms could gain importance with the use of appropriate preconditioners in this regard. However, existing preconditioners proposed for interior-point methods for nonlinear problems perform better than their counterparts for linear problems. Future research should investigate ways of how to overcome this deficiency, allowing the use of SIPM even for a saddle point matrix $\tilde{\mathcal{S}}$ of large dimension.

Further adjustments and modifications of the accelerated L-shaped method are attractive challenges for future work. Especially its application to an even broader problem class appears promising. For instance, if the master problem cannot be solved by a commercial solver due to its large size, alternative methods like the one-tree approach outlined in Section 4.3 may be employed.

Another future research direction is the need for more realistic case studies. The humanitarian literature is scarce in this regard, leading to unrealistic and misleading results with limited information value. In Section 4.1.3 the extreme differences concerning location decisions between the case study of Rawls and Turnquist [2010] and the newly proposed large-scale case study were highlighted. Some facilities suggested by Rawls and Turnquist [2010] are located in areas which can be classified as vulnerable if all historical hurricanes are taken into account. It can be concluded that results based on illustrative examples and unrealistic data are of no practical use. Especially defining realistic scenarios is decisive to gain valuable and applicable insights and to derive recommendations for relief organizations. In Chapter 4 a first step in this direction was made by designing two large-scale case studies based on more realistic data.

The newly developed solution method enables relief agencies to integrate two-stage stochastic programs into practice, especially in emergency situations. Such optimization models can lead to better justified decisions, more responsible handling of donations, more satisfied demand and to reduced numbers of fatalities.

Bibliography

AccuWeather. AccuWeather predicts economic cost of Harvey, Irma to be 290 billion, 2017. URL https://www.accuweather.com/en/weather-news/accuweather-predicts-economic-cost-of-harvey-irma-to-be-290-billion/70002686. Accessed: 2017-10-10.

C. Adinolfi, D. Bassiouni, H. Lauritzsen, and H. Williams. Humanitarian Response Review. Technical report, United Nations Emergency Relief Coordinator and Office for the Coordination of Humanitarian Affairs, 2005.

M. Ahmadi, A. Seifi, and B. Tootooni. A humanitarian logistics model for disaster relief operation considering network failure and standard relief time: A case study on San Francisco district. *Transportation Research Part E: Logistics and Transportation Review*, 75:145–163, 2015. ISSN 1366-5545. doi: http://dx.doi.org/10.1016/j.tre.2015.01.008. URL http://www.sciencedirect.com/science/article/pii/S1366554515000095.

G. Al-Jeiroudi, J. Gondzio, and J. Hall. Preconditioning indefinite systems in interior point methods for large scale linear optimisation. *Optimization Methods and Software*, 23(3):345–363, 2008. doi: 10.1080/10556780701535910. URL http://dx.doi.org/10.1080/10556780701535910.

D. Alem, A. Clark, and A. Moreno. Stochastic network models for logistics planning in disaster relief. *European Journal of Operational Research*, 255(1):187–206, 2016. ISSN 0377-2217. doi: http://dx.doi.org/10.1016/j.ejor.2016.04.041. URL http://www.sciencedirect.com/science/article/pii/S0377221716302788.

D. Alexander. *Principles of Emergency Planning and Management*. Oxford University Press, Oxford, 2002. ISBN 9780195218381. URL https://books.google.de/books?id=iLqMSDgecHQC.

A. K. Andreas and J. C. Smith. Decomposition algorithms for the design of a nonsimultaneous capacitated evacuation tree network. *Networks*, 53(2):91–103, 2009. ISSN 1097-0037. doi: 10.1002/net.20278. URL http://dx.doi.org/10.1002/net.20278.

AOML. The Atlantic Oceanographic and Meteorological Laboratory: How many tropical cyclones have there been each year in the Atlantic basin? What years were the greatest and fewest seen?, 2016. URL http://www.aoml.noaa.gov/hrd/tcfaq/E11.html. Accessed: 2017-06-20.

R. K. Arora. *Optimization*. Taylor & Francis Inc, Boca Raton, 2015. ISBN 978-1-4987-2112-7.

E. Graß, *An Accelerated Solution Method for Two-Stage Stochastic Models in Disaster Management*, Mathematische Optimierung und Wirtschaftsmathematik | Mathematical Optimization and Economathematics,
https://doi.org/10.1007/978-3-658-24081-3

X. Bai. Two-Stage Multiobjective Optimization for Emergency Supplies Allocation Problem under Integrated Uncertainty. *Mathematical Problems in Engineering*, 2016:1–13, 2016. doi: 10.1155/2016/2823835. URL http://dx.doi.org/10.1155/2016/2823835.

B. Balcik and B. M. Beamon. Facility location in humanitarian relief. *International Journal of Logistics Research and Applications*, 11(2):101–121, 2008. doi: 10.1080/13675560701561789. URL http://www.tandfonline.com/doi/abs/10.1080/13675560701561789.

B. Balcik, B. M. Beamon, C. C. Krejci, K. M. Muramatsu, and M. Ramirez. Coordination in humanitarian relief chains: Practices, challenges and opportunities. *International Journal of Production Economics*, 126(1):22–34, 2010. ISSN 0925-5273. doi: https://doi.org/10.1016/j.ijpe.2009.09.008. URL http://www.sciencedirect.com/science/article/pii/S092552730900365X.

V. A. Bañuls, M. Turoff, and S. R. Hiltz. Supporting collaborative scenario analysis through cross-impact. In *Proceedings of the 9th International Conference on Information Systems for Crisis Response and Management*, 2012.

G. Barbarosoğlu and Y. Arda. A two-stage stochastic programming framework for transportation planning in disaster response. *Journal of the Operational Research Society*, 55(1):43–53, 2004. ISSN 1476-9360. doi: 10.1057/palgrave.jors.2601652. URL http://dx.doi.org/10.1057/palgrave.jors.2601652.

M. Benzi, G. H. Golub, and J. Liesen. Numerical solution of saddle point problems. *Acta Numerica*, 14:1–137, 2005. doi: 10.1017/S0962492904000212.

J. Birge and F. Louveaux. *Introduction to Stochastic Programming*. Springer Science & Business Media, New York, 2011.

S. Bocanegra, F. F. Campos, and A. R. L. Oliveira. Using a hybrid preconditioner for solving large-scale linear systems arising from interior point methods. *Computational Optimization and Applications*, 36(2):149–164, 2007. ISSN 1573-2894. doi: 10.1007/s10589-006-9009-5. URL http://dx.doi.org/10.1007/s10589-006-9009-5.

S. Bocanegra, J. Castro, and A. R. Oliveira. Improving an interior-point approach for large block-angular problems by hybrid preconditioners. *European Journal of Operational Research*, 231(2):263–273, 2013. ISSN 0377-2217. doi: http://dx.doi.org/10.1016/j.ejor.2013.04.007. URL http://www.sciencedirect.com/science/article/pii/S0377221713003056.

S. Boyd and L. Vandenberghe. *Convex Optimization*. Cambridge University Press, 2004. ISBN 978-0521833783. URL http://www.ebook.de/de/product/3677442/stephen_stanford_university_california_boyd_lieven_university_of_california_los_angeles_vandenberghe_convex_optimization.html.

A. Bozorgi-Amiri, M. S. Jabalameli, M. Alinaghian, and M. Heydari. A modified particle swarm optimization for disaster relief logistics under uncertain environment. *The International Journal of Advanced Manufacturing Technology*, 60(1):357–371, 2012. ISSN 1433-3015. doi: 10.1007/s00170-011-3596-8. URL `http://dx.doi.org/10.1007/s00170-011-3596-8`.

A. Bozorgi-Amiri, M. Jabalameli, and S. A. e Hashem. A multi-objective robust stochastic programming model for disaster relief logistics under uncertainty. *OR spectrum*, 35(4): 905–933, 2013.

Y. Cai, G. Huang, Q. Tan, and B. Chen. Identification of optimal strategies for improving eco-resilience to floods in ecologically vulnerable regions of a wetland. *Ecological Modelling*, 222(2):360–369, 2011. ISSN 0304-3800. doi: http://doi.org/10.1016/j.ecolmodel.2009.12.012. URL `http://www.sciencedirect.com/science/article/pii/S0304380009008424`.

A. M. Campbell and P. C. Jones. Prepositioning supplies in preparation for disasters. *European Journal of Operational Research*, 209(2):156–165, 2011. ISSN 0377-2217. doi: http://doi.org/10.1016/j.ejor.2010.08.029. URL `http://www.sciencedirect.com/science/article/pii/S0377221710005898`.

J. Castro. A Specialized Interior-Point Algorithm for Multicommodity Network Flows. *SIAM Journal on Optimization*, 10(3):852–877, 2000. doi: 10.1137/S1052623498341879. URL `https://doi.org/10.1137/S1052623498341879`.

A. Caunhye, X. Nie, and S. Pokharel. Optimization models in emergency logistics: A literature review. *Socio-Economic Planning Sciences*, 46(1):4–13, 2012.

A. M. Caunhye, Y. Zhang, M. Li, and X. Nie. A location-routing model for prepositioning and distributing emergency supplies. *Transportation Research Part E: Logistics and Transportation Review*, 90:161–176, 2016. ISSN 1366-5545. doi: http://doi.org/10.1016/j.tre.2015.10.011. URL `http://www.sciencedirect.com/science/article/pii/S1366554515002033`.

CB. United States Census Bureau, 2016. URL `https://www.census.gov/data/tables/2016/demo/popest/counties\-total.html`. Accessed: 2017-06-22.

M. Chang, Y. Tseng, and J. Chen. A scenario planning approach for the flood emergency logistics preparation problem under uncertainty. *Transportation Research Part E: Logistics and Transportation Review*, 43(6):737–754, 2007.

T. Comes, N. Wijngaards, and F. Schultmann. Efficient scenario updating in emergency management. In *Proceedings of the 9th International Conference on Information Systems for Crisis Response and Management*, 2012.

D. P. Coppola. *International Disaster Management*. Butterworth-Heinemann, Amsterdam, second edition edition, 2011. ISBN 978-0-12-382174-4.

CRED/ UNISDR. The Human Cost of Weather related Disasters, 2015. URL https://www.unisdr.org/2015/docs/climatechange/. Accessed: 2017-06-08.

L. B. Davis, F. Samanlioglu, X. Qu, and S. Root. Inventory planning and coordination in disaster relief efforts. *International Journal of Production Economics*, 141(2):561–573, 2013. ISSN 0925-5273. doi: http://doi.org/10.1016/j.ijpe.2012.09.012. URL http://www.sciencedirect.com/science/article/pii/S0925527312004082.

J. Desrosiers and M. E. Lübbecke. *A Primer in Column Generation*, pages 1–32. Springer US, Boston, MA, 2005. ISBN 978-0-387-25486-9. doi: 10.1007/0-387-25486-2_1. URL https://doi.org/10.1007/0-387-25486-2_1.

L. Dignan. Tricky currents; tsunami relief is a challenge when supply chains are blocked by cows and roads don't exist. *Baseline*, 1(39):30, 2005.

Direct Relief. Hurricane preparedness program, 2017. URL https://www.directrelief.org/emergency/. Accessed: 2017-08-18.

U. Diwekar. *Introduction to Applied Optimization*. SPRINGER VERLAG GMBH, 2008. ISBN 978-0-387-76634-8. URL http://www.ebook.de/de/product/7522360/urmila_diwekar_introduction_to_applied_optimization.html.

A. Döyen, N. Aras, and G. Barbarosoğlu. A two-echelon stochastic facility location model for humanitarian relief logistics. *Optimization Letters*, 6(6):1123–1145, 2012. ISSN 1862-4480. doi: 10.1007/s11590-011-0421-0. URL http://dx.doi.org/10.1007/s11590-011-0421-0.

J. Dupačová, G. Consigli, and S. W. Wallace. Scenarios for Multistage Stochastic Programs. *Annals of Operations Research*, 100(1):25–53, 2000. ISSN 1572-9338. doi: 10.1023/A:1019206915174. URL http://dx.doi.org/10.1023/A:1019206915174.

S. Duran, M. Gutierrez, and P. Keskinocak. Pre-positioning of emergency items for CARE international. *Interfaces*, 41(3):223–237, 2011.

M. Ehrgott. *Multicriteria Optimization*. Springer, Berlin; Heidelberg, 2005. ISBN 978-3-540-21398-7.

J. Elliott, J. Eisinger, and L. Sullivan. The Red Cross' Secret Disaster. Technical report, ProPublica, 2014.

EM-DAT. Advanced Search - The International Disaster Database, 2017. URL http://www.emdat.be/advancedsearch/index.html. Accessed: 2017-08-18.

Environment Agency. Flood Forecasts, 2017. URL http://apps.environment-agency.gov.uk/flood/3days/125305.aspx. Accessed: 2017-11-24.

M. Falasca and C. W. Zobel. A two-stage procurement model for humanitarian relief supply chains. *Journal of Humanitarian Logistics and Supply Chain Management*, 1(2): 151–169, 2011. doi: 10.1108/20426741111188329. URL http://dx.doi.org/10.1108/20426741111188329.

Y. Fan and C. Liu. Solving Stochastic Transportation Network Protection Problems Using the Progressive Hedging-based Method. *Networks and Spatial Economics*, 10 (2):193–208, 2010. ISSN 1572-9427. doi: 10.1007/s11067-008-9062-y. URL http://dx.doi.org/10.1007/s11067-008-9062-y.

M. Ferris, O. Mangasarian, and S. Wright. *Linear Programming with MATLAB*. Society for Industrial and Applied Mathematics, Philadelphia, 2007. doi: 10.1137/1.9780898718775. URL http://epubs.siam.org/doi/abs/10.1137/1.9780898718775.

G. Galindo and R. Batta. Prepositioning of supplies in preparation for a hurricane under potential destruction of prepositioned supplies. *Socio-Economic Planning Sciences*, 47 (1):20–37, 2013. ISSN 0038-0121. doi: http://doi.org/10.1016/j.seps.2012.11.002. URL http://www.sciencedirect.com/science/article/pii/S0038012112000596.

F. W. Glover and M. Laguna. *Tabu Search*. Springer, Boston, 1997 edition, 1997. ISBN 978-0-7923-9965-0.

G. H. Golub and C. F. Van Loan. *Matrix Computations*. John Hopkins, Baltimore, 4 th edition, 2013. ISBN 978-4214-0794-4.

J. Gondzio. Interior point methods 25 years later. *European Journal of Operational Research*, 218(3):587–601, 2012. ISSN 0377-2217. doi: http://doi.org/10.1016/j.ejor.2011.09.017. URL http://www.sciencedirect.com/science/article/pii/S0377221711008204.

E. Grass. Additonal material for the dissertation Än Accelerated Solution Method for Two-Stage Stochastic Models in Disaster Management": Data, MATLAB codes and results, Dec. 2017. URL https://doi.org/10.5281/zenodo.1012684.

E. Grass and K. Fischer. *Prepositioning of Relief Items Under Uncertainty: A Classification of Modeling and Solution Approaches for Disaster Management*, pages 189–202. Springer International Publishing, Cham, 2016a. ISBN 978-3-319-20863-3. doi: 10.1007/978-3-319-20863-3_14. URL https://doi.org/10.1007/978-3-319-20863-3_14.

E. Grass and K. Fischer. Two-stage stochastic programming in disaster management: A literature survey. *Surveys in Operations Research and Management Science*, 21(2): 85–100, 2016b. ISSN 1876-7354. doi: http://dx.doi.org/10.1016/j.sorms.2016.11.002. URL //www.sciencedirect.com/science/article/pii/S1876735416300162.

E. Grass, K. Fischer, and A. Rams. An accelerated l-shaped method for solving two-stage stochastic programs in disaster management. *Annals of Operations Research*, pages 1–26, 2018.

D. Guha-Sapir, F. Vos, R. Below, and S. Ponserre. Annual disaster statistical review 2011: The numbers and trends. Brussels: Centre for Research on the Epidemiology of Disasters, 2012.

Gurobi. Gurobi: Parameter Documentation, 2017. URL http://www.gurobi.com/documentation/7.5/refman/method.html#parameter:Method. Accessed: 2017-12-05.

Gurobi. Gurobi Optimization: Documentation, 2017a. URL https://www.gurobi.com/documentation/7.0/refman/mipgap2.html. Accessed: 2017-07-09.

Gurobi. Gurobi Optimization: Mixed-Integer Programming, 2017b. URL {http://www.gurobi.com/resources/getting-started/mip-basics}. Accessed: 2017-09-13.

W. J. Gutjahr and P. C. Nolz. Multicriteria optimization in humanitarian aid. *European Journal of Operational Research*, 252(2):351–366, 2016. ISSN 0377-2217. doi: http://doi.org/10.1016/j.ejor.2015.12.035. URL http://www.sciencedirect.com/science/article/pii/S0377221715011741.

J. L. Higle. *Stochastic Programming: Optimization When Uncertainty Matters*, chapter Chapter 2, pages 30–53. Informs, 2006. doi: 10.1287/educ.1053.0016. URL http://pubsonline.informs.org/doi/abs/10.1287/educ.1053.0016.

X. Hong, M. A. Lejeune, and N. Noyan. Stochastic network design for disaster preparedness. *IIE Transactions*, 47(4):329–357, 2015. doi: 10.1080/0740817X.2014.919044. URL http://dx.doi.org/10.1080/0740817X.2014.919044.

R. A. Horn and C. R. Johnson. *Matrix Analysis*. Cambridge University Press, 2012. ISBN 978-0-521-83940-2. URL http://www.ebook.de/de/product/19399171/roger_a_horn_charles_r_johnson_matrix_analysis.html.

M. C. Hoyos, R. S. Morales, and R. Akhavan-Tabatabaei. OR models with stochastic components in disaster operations management: A literature survey. *Computers & Industrial Engineering*, 82:183–197, 2015. ISSN 0360-8352. doi: http://doi.org/10.1016/j.cie.2014.11.025. URL http://www.sciencedirect.com/science/article/pii/S0360835214004136.

S.-L. Hu, C.-F. Han, and L.-P. Meng. A scenario planning approach for propositioning rescue centers for urban waterlog disasters. *Computers & Industrial Engineering*, 87: 425–435, 2015. ISSN 0360-8352. doi: http://doi.org/10.1016/j.cie.2015.05.036. URL http://www.sciencedirect.com/science/article/pii/S0360835215002594.

Y. Hu, K. Maguire, and R. Blake. A multilevel unsymmetric matrix ordering algorithm for parallel process simulation. *Computers & Chemical Engineering*, 23(11):1631 – 1647, 2000. ISSN 0098-1354. doi: http://dx.doi.org/10.1016/S0098-1354(99)00314-2. URL http://www.sciencedirect.com/science/article/pii/S0098135499003142.

IFRC. International Federation of Red Cross and Red Crescent Societies-Items Catalogue, 2016. URL `http://procurement.ifrc.org/catalogue/detail.aspx`. Accessed: 2017-01-26.

ISCRAM. Private Communication at the International Conference on Information Systems for Crisis Response and Management, 2017.

C. Keller, N. I. M. Gould, and A. J. Wathen. Constraint Preconditioning for Indefinite Linear Systems. *SIAM Journal on Matrix Analysis and Applications*, 21(4): 1300–1317, 2000. doi: 10.1137/S0895479899351805. URL `https://doi.org/10.1137/S0895479899351805`.

W. Klibi, S. Ichoua, and A. Martel. *Prepositioning emergency supplies to support disaster relief: a stochastic programming approach.* Faculté des sciences de l'administration, Université Laval, Québec, 2013.

R. D. Knabb, J. R. Rhome, and D. P. Brown. Tropical Cyclone Report Hurricane Katrina 23-30 August 2005. Technical report, NHC, 2011.

G. Kovács and K. M. Spens. Humanitarian logistics in disaster relief operations. *International Journal of Physical Distribution & Logistics Management*, 37(2):99–114, 2007. doi: 10.1108/09600030710734820. URL `http://dx.doi.org/10.1108/09600030710734820`.

R. H. Kwon. *Introduction to linear optimization and extensions with MATLAB.* CRC Press, Boca Raton, 2013. ISBN 978-1-4398-6263-6.

C. Landsea. NOAA Communications & External Affairs: hurricane probability calculations. Private Communication, Dec. 2017.

G. Laporte and F. V. Louveaux. The integer L-shaped method for stochastic integer programs with complete recourse. *Operations Research Letters*, 13(3):133–142, 1993. ISSN 0167-6377. doi: http://dx.doi.org/10.1016/0167-6377(93)90002-X. URL `http://www.sciencedirect.com/science/article/pii/016763779390002X`.

A. C. Li, L. Nozick, N. Xu, and R. Davidson. Shelter location and transportation planning under hurricane conditions. *Transportation Research Part E: Logistics and Transportation Review*, 48(4):715–729, 2012. ISSN 1366-5545. doi: http://dx.doi.org/10.1016/j.tre.2011.12.004. URL `http://www.sciencedirect.com/science/article/pii/S1366554511001566`.

L. Li, M. Jin, and L. Zhang. Sheltering network planning and management with a case in the Gulf Coast region. *International Journal of Production Economics*, 131(2):431–440, 2011. ISSN 0925-5273. doi: http://dx.doi.org/10.1016/j.ijpe.2010.12.013. URL `http://www.sciencedirect.com/science/article/pii/S0925527310004822`.

C. Liu, Y. Fan, and F. Ordóñez. A two-stage stochastic programming model for transportation network protection. *Computers & Operations Research*, 36(5):1582–1590, 2009. ISSN 0305-0548. doi: http://dx.doi.org/10.1016/j.cor.2008.03.001. URL `http://www.sciencedirect.com/science/article/pii/S030505480800052X`.

Z. Liu and G. Huang. Dual-Interval Two-Stage Optimization for Flood Management and Risk Analyses. *Water Resources Management*, 23(11):2141–2162, 2009. ISSN 1573-1650. doi: 10.1007/s11269-008-9375-0. URL http://dx.doi.org/10.1007/s11269-008-9375-0.

E. J. J. Lodree, K. N. Ballard, and C. H. Song. Pre-positioning hurricane supplies in a commercial supply chain. *Socio-Economic Planning Sciences*, 46(4):291–305, 2012. ISSN 0038-0121. doi: http://dx.doi.org/10.1016/j.seps.2012.03.003. URL http://www.sciencedirect.com/science/article/pii/S0038012112000183.

V. Maniezzo, T. Stützle, and S. Voß. *Matheuristics: Hybridizing Metaheuristics and Mathematical Programming*. 10. Springer US, New York, 2010. ISBN 978-1-4419-1306-7. doi: 10.1007/978-1-4419-1306-7.

I. Maqsood and G. H. Huang. A dual two-stage stochastic model for flood management with inexact-integer analysis under multiple uncertainties. *Stochastic Environmental Research and Risk Assessment*, 27(3):643–657, 2013. ISSN 1436-3259. doi: 10.1007/s00477-012-0629-2. URL http://dx.doi.org/10.1007/s00477-012-0629-2.

T. Mathew. *Domain Decomposition Methods for the Numerical Solution of Partial Differential Equations*. Springer Berlin Heidelberg, 2008. ISBN 978-3-540-77209-5. URL http://www.ebook.de/de/product/8897526/tarek_mathew_domain_decomposition_methods_for_the_numerical_solution_of_partial_differential_equations.html.

H. O. Mete and Z. B. Zabinsky. Stochastic optimization of medical supply location and distribution in disaster management. *International Journal of Production Economics*, 126(1):76–84, 2010. ISSN 0925-5273. doi: http://doi.org/10.1016/j.ijpe.2009.10.004. URL http://www.sciencedirect.com/science/article/pii/S0925527309003582.

A. Mitsotakis and G. Kassaras. Managing disaster in the Ionian Sea: Planning and optimizing logistics for disaster relief operations for the island of Kefalonia. Technical report, DTIC Document, 2010.

S. Moore, E. Eng, and M. Daniel. International NGOs and the role of network centrality in humanitarian aid operations: a case study of coordination during the 2000 Mozambique floods. *Disasters*, 27(4):305–318, 2003. ISSN 0361-3666.

MSF. Ärzte ohne Grenzen e.V.-Private Communication, 2016.

P. Murali, F. Ordóñez, and M. M. Dessouky. Facility location under demand uncertainty: Response to a large-scale bio-terror attack. *Socio-Economic Planning Sciences*, 46(1): 78–87, 2012. ISSN 0038-0121. doi: http://doi.org/10.1016/j.seps.2011.09.001. URL http://www.sciencedirect.com/science/article/pii/S0038012111000450.

J. Murphy. *Benders, Nested Benders and Stochastic Programming: An Intuitive Introduction*. CUED/F-INFENG/TR. University of Cambridge, Department of Engineering, 2013. URL https://books.google.de/books?id=BZvUrQEACAAJ.

J. Naoum-Sawaya and S. Elhedhli. An interior-point Benders based branch-and-cut algorithm for mixed integer programs. *Annals of Operations Research*, 210(1):33–55, 2013. ISSN 1572-9338. doi: 10.1007/s10479-010-0806-y. URL http://dx.doi.org/10.1007/s10479-010-0806-y.

M. Ng and S. Waller. The evacuation optimal network design problem: model formulation and comparisons. *Transportation Letters*, 1(2):111–119, 2009. doi: 10.3328/TL.2009.01.02.111-119. URL http://dx.doi.org/10.3328/TL.2009.01.02.111-119.

NHC. National Hurricane Center, 2005. URL http://www.nhc.noaa.gov/archive/2005/KATRINA_graphics.shtml. Accessed: 2017-06-22.

NHC. National Hurricane Center: Tropical Cyclone Wind Speed Probabilities Products, 2014. URL http://www.nhc.noaa.gov/about/pdf/About_Windspeed_Probabilities.pdf. Accessed: 2017-11-27.

NHC. National Hurricane Center, 2017a. URL http://www.nhc.noaa.gov/. Accessed: 2017-08-18.

NHC. National Hurricane Center: Saffir-Simpson Hurricane Wind Scale, 2017b. URL http://www.nhc.noaa.gov/aboutsshws.php. Accessed: 2017-07-05.

NHCFP. National Hurricane Center Forecast Process, 2017. URL http://www.hurricanescience.org/science/forecast/forecasting/forecastprocess/. Accessed: 2017-06-23.

NOAA and AOML. National Oceanic and Atmospheric Administration/Atlantic Oceanographic and Meteorological Laboratory, 2016. URL http://www.aoml.noaa.gov/hrd/tcfaq/D5.html. Accessed: 2017-08-18.

J. Nocedal and S. J. Wright. *Numerical Optimization*. Springer, New York, 2nd edition, 2006.

N. Noyan. Risk-averse two-stage stochastic programming with an application to disaster management. *Computers & Operations Research*, 39(3):541–559, 2012. ISSN 0305-0548. doi: http://doi.org/10.1016/j.cor.2011.03.017. URL http://www.sciencedirect.com/science/article/pii/S0305054811000931.

N. Noyan, B. Balcik, and S. Atakan. A Stochastic Optimization Model for Designing Last Mile Relief Networks. *Transportation Science*, 50(3):1092–1113, 2016. doi: 10.1287/trsc.2015.0621. URL http://dx.doi.org/10.1287/trsc.2015.0621.

M. T. Ortuño, P. Cristóbal, J. M. Ferrer, F. J. Martín-Campo, S. Muñoz, G. Tirado, and B. Vitoriano. *Decision Aid Models and Systems for Humanitarian Logistics. A Survey*, pages 17–44. Atlantis Press, Paris, 2013. ISBN 978-94-91216-74-9. doi: 10.2991/978-94-91216-74-9-2. URL http://dx.doi.org/10.2991/978-94-91216-74-9-2.

PAHO. *Humanitarian Supply Management and Logistics in the Health Sector.* Pan American Health Organization and World Health Organization, Washington, D.C., 2001.

S. Peeta, F. S. Salman, D. Gunnec, and K. Viswanath. Pre-disaster investment decisions for strengthening a highway network. *Computers & Operations Research*, 37(10):1708–1719, 2010. ISSN 0305-0548. doi: http://dx.doi.org/10.1016/j.cor.2009.12.006. URL `http://www.sciencedirect.com/science/article/pii/S0305054809003311`.

A. M. Quarteroni, R. Sacco, and F. Saleri. *Numerical Mathematics.* Springer-Verlag GmbH, 2006. ISBN 978-3540346586. URL `http://www.ebook.de/de/product/5810163/alfio_m_quarteroni_riccardo_sacco_fausto_saleri_numerical_mathematics.html`.

R. Rahmaniani, T. G. Crainic, M. Gendreau, and W. Rei. The Benders decomposition algorithm: A literature review. *European Journal of Operational Research*, 259(3): 801–817, 2017. ISSN 0377-2217. doi: http://dx.doi.org/10.1016/j.ejor.2016.12.005. URL `www.sciencedirect.com/science/article/pii/S0377221716310244`.

A. Rams. Vorpositionierung von Hilfsgütern zur Katastrophenvorsorge unter Berücksichtigung zeitkritischer Aspekte. Master's thesis, 2016.

S. Rath, M. Gendreau, and W. J. Gutjahr. Bi-objective stochastic programming models for determining depot locations in disaster relief operations. *International Transactions in Operational Research*, 23(6):997–1023, 2016. ISSN 1475-3995. doi: 10.1111/itor.12163. URL `http://dx.doi.org/10.1111/itor.12163`.

C. G. Rawls and M. A. Turnquist. Pre-positioning of emergency supplies for disaster response. *Transportation Research Part B: Methodological*, 44(4):521–534, 2010. ISSN 0191-2615. doi: http://dx.doi.org/10.1016/j.trb.2009.08.003. URL `http://www.sciencedirect.com/science/article/pii/S0191261509001118`.

C. G. Rawls and M. A. Turnquist. Pre-positioning planning for emergency response with service quality constraints. *OR Spectrum*, 33(3):481–498, 2011. ISSN 1436-6304. doi: 10.1007/s00291-011-0248-1. URL `http://dx.doi.org/10.1007/s00291-011-0248-1`.

C. G. Rawls and M. A. Turnquist. Pre-positioning and dynamic delivery planning for short-term response following a natural disaster. *Socio-Economic Planning Sciences*, 46(1): 46–54, 2012. ISSN 0038-0121. doi: http://dx.doi.org/10.1016/j.seps.2011.10.002. URL `http://www.sciencedirect.com/science/article/pii/S0038012111000498`. Special Issue: Disaster Planning and Logistics: Part 1.

RCRC. *Emergency Items Catalogue*, volume 1. Red Cross Red Crescent, 3 edition, 2009.

C. Renkli and S. Duran. Pre-positioning disaster response facilities and relief items. *Human and Ecological Risk Assessment: An International Journal*, 21(5):1169–1185, 2015. doi: 10.1080/10807039.2014.957940. URL `http://dx.doi.org/10.1080/10807039.2014.957940`.

S. J. Rennemo, K. F. Rø, L. M. Hvattum, and G. Tirado. A three-stage stochastic facility routing model for disaster response planning. *Transportation Research Part E: Logistics and Transportation Review*, 62:116–135, 2014. ISSN 1366-5545. doi: http://doi.org/10.1016/j.tre.2013.12.006. URL `http://www.sciencedirect.com/science/article/pii/S1366554513002007`.

M. Rezaei-Malek, R. Tavakkoli-Moghaddam, N. Cheikhrouhou, and A. Taheri-Moghaddam. An approximation approach to a trade-off among efficiency, efficacy, and balance for relief pre-positioning in disaster management. *Transportation Research Part E: Logistics and Transportation Review*, 93:485–509, 2016. ISSN 1366-5545. doi: http://dx.doi.org/10.1016/j.tre.2016.07.003. URL `http://www.sciencedirect.com/science/article/pii/S136655451630134X`.

R. T. Rockafellar and R. J.-B. Wets. Scenarios and Policy Aggregation in Optimization Under Uncertainty. *Mathematics of Operations Research*, 16(1):119–147, 1991. doi: 10.1287/moor.16.1.119. URL `http://dx.doi.org/10.1287/moor.16.1.119`.

B. Rottkemper and K. Fischer. Decision making in humanitarian logistics–A multi-objective optimization model for relocating relief goods during disaster recovery operations. In *Proceedings of the 10th international conference on information systems for crisis response management, Germany*, pages 647–657, 2013. URL `http://ww.iscram.org/legacy/ISCRAM2013/files/158.pdf`.

P. Rubin. Benders decomposition then and now, 2011. URL `http://orinanobworld.blogspot.de/2011/10/benders-decomposition-then-and-now.html`.

Y. Saad. *Iterative methods for sparse linear systems.* Society for Industrial and Applied Mathematics, Philadelphia, 2 edition, 2003. URL `http://epubs.siam.org/doi/pdf/10.1137/1.9780898718003.bm`.

J. Salmerón and A. Apte. Stochastic Optimization for Natural Disaster Asset Prepositioning. *Production and Operations Management*, 19(5):561–574, 2010. ISSN 1937-5956. doi: 10.1111/j.1937-5956.2009.01119.x. URL `http://dx.doi.org/10.1111/j.1937-5956.2009.01119.x`.

SCEMD. South Carolina Emergency Management Division-South Carolina Logistical Operations Plan: Appendix 7, 2007. URL `http://dc.statelibrary.sc.gov/handle/10827/20614`. Accessed: 2017-08-18.

K. Schade. *Stochastische Optimierung.* Vieweg+Teubner Verlag, Wiesbaden, 2012. ISBN 978-3-8348-8345-2.

S. P. Schnaars. How to develop and use scenarios. *Long Range Planning*, 20(1):105 – 114, 1987. ISSN 0024-6301. doi: https://doi.org/10.1016/0024-6301(87)90038-0. URL `http://www.sciencedirect.com/science/article/pii/0024630187900380`.

A. Shapiro and A. Philpott. A tutorial on stochastic programming, manuscript, 2007, 2007. URL `http://www2.isye.gatech.edu/ashapiro/publications.html`. Accessed: 2017-08-14.

Z. Shen, M. M. Dessouky, and F. Ordóñez. A two-stage vehicle routing model for large-scale bioterrorism emergencies. *Networks*, 54(4):255–269, 2009. ISSN 1097-0037. doi: 10.1002/net.20337. URL `http://dx.doi.org/10.1002/net.20337`.

T. M. Simatupang, A. C. Wright, and R. Sridharan. The knowledge of coordination for supply chain integration. *Business Process Management Journal*, 8(3):289–308, 2002. doi: 10.1108/14637150210428989. URL `http://dx.doi.org/10.1108/14637150210428989`.

R. M. V. Slyke and R. Wets. L-Shaped Linear Programs with Applications to Optimal Control and Stochastic Programming. *SIAM Journal on Applied Mathematics*, 17(4): 638–663, 1969. doi: 10.1137/0117061. URL `http://dx.doi.org/10.1137/0117061`.

The New York Times. Reports Critique U.S. Red Cross's Katrina Response, 2006. URL `http://www.nytimes.com/2006/04/04/us/reports-critique-us-red-crosss-katrina-response.html?mcubz=3`. Accessed: 2017-14-09.

S. Tofighi, S. Torabi, and S. Mansouri. Humanitarian logistics network design under mixed uncertainty. *European Journal of Operational Research*, 250(1):239–250, 2016. ISSN 0377-2217. doi: http://dx.doi.org/10.1016/j.ejor.2015.08.059. URL `http://www.sciencedirect.com/science/article/pii/S0377221715008152`.

R. Tomasini and L. van Wassenhove. *Humanitarian Logistics*. Palgrave Macmillan, Houndmills, Basingstoke; New York, 2009. ISBN 978-0-230-23348-5.

F. Toyasaki and T. Wakolbinger. Impacts of earmarked private donations for disaster fundraising. *Annals of Operations Research*, 221(1):427–447, 2014. ISSN 1572-9338. doi: 10.1007/s10479-011-1038-5. URL `http://dx.doi.org/10.1007/s10479-011-1038-5`.

F. Tricoire, A. Graf, and W. J. Gutjahr. The bi-objective stochastic covering tour problem. *Computers & Operations Research*, 39(7):1582–1592, 2012. ISSN 0305-0548. doi: http://dx.doi.org/10.1016/j.cor.2011.09.009. URL `http://www.sciencedirect.com/science/article/pii/S030505481100270X`.

URI and GSO. University of Rhode Island and Graduate School of Oceanography-Hurricanes: Science and Society, 2016. URL `http://www.hurricanescience.org/science/science/hurricanestructure/`. Accessed: 2017-08-18.

L. N. Van Wassenhove. Humanitarian aid logistics: supply chain management in high gear. *Journal of the Operational Research Society*, 57(5):475–489, 2006. ISSN 1476-9360. doi: 10.1057/palgrave.jors.2602125. URL `http://dx.doi.org/10.1057/palgrave.jors.2602125`.

R. J. Vanderbei. *Linear Programming*. Springer US, New York, 2014. doi: 10.1007/978-1-4614-7630-6.

A. Verma and G. M. Gaukler. Pre-positioning disaster response facilities at safe locations: An evaluation of deterministic and stochastic modeling approaches. *Computers & Operations Research*, 62:197–209, 2015. ISSN 0305-0548. doi: http://dx.doi.org/10.1016/j.cor.2014.10.006. URL http://www.sciencedirect.com/science/article/pii/S0305054814002639.

D. W. Walkup and R. J.-B. Wets. Stochastic Programs with Recourse. *SIAM Journal on Applied Mathematics*, 15(5):1299–1314, 1967. doi: 10.1137/0115113. URL https://doi.org/10.1137/0115113.

J. Wang, H. Yang, and J. Zhu. A Two-stage Stochastic Programming Model for Emergency Resources Storage Region Division. *Systems Engineering Procedia*, 5:125–130, 2012. ISSN 2211-3819. doi: http://dx.doi.org/10.1016/j.sepro.2012.04.020. URL http://www.sciencedirect.com/science/article/pii/S221138191200063X.

S. Wang and G. Huang. A two-stage mixed-integer fuzzy programming with interval-valued membership functions approach for flood-diversion planning. *Journal of Environmental Management*, 117:208–218, 2013. ISSN 0301-4797. doi: http://dx.doi.org/10.1016/j.jenvman.2012.12.037. URL http://www.sciencedirect.com/science/article/pii/S0301479713000078.

WFP. World Food Programme, 2017. URL https://www.wfp.org/logistics/humanitarian-response-depot. Accessed: 2017-08-18.

S. J. Wright. *Primal-Dual Interior-Point Methods.* Society for Industrial and Applied Mathematics, Philadelphia, 1997. doi: 10.1137/1.9781611971453.bm.

H. Zhang and D. Liu. *Fuzzy Modeling and Fuzzy Control.* Birkhäuser, Boston; Basel; Berlin, 2006. ISBN 978-0-8176-4491-8.

Y. Zhang. Solving large-scale linear programs by interior-point methods under the Matlab Environment. *Optimization Methods and Software*, 10(1):1–31, 1998. doi: 10.1080/10556789808805699. URL http://dx.doi.org/10.1080/10556789808805699.

M. R. Zolfaghari and E. Peyghaleh. Implementation of Equity in Resource Allocation for Regional Earthquake Risk Mitigation Using Two-Stage Stochastic Programming. *Risk Analysis*, 35(3):434–458, 2015. ISSN 1539-6924. URL http://dx.doi.org/10.1111/risa.12321.

L. Özdamar and M. A. Ertem. Models, solutions and enabling technologies in humanitarian logistics. *European Journal of Operational Research*, 244(1):55–65, 2015. ISSN 0377-2217. doi: https://doi.org/10.1016/j.ejor.2014.11.030. URL http://www.sciencedirect.com/science/article/pii/S0377221714009539.

A Appendix

A.1 The Recourse Function: An Example

The following example of a piece-wise and convex recourse function $Q(\chi, \xi(s))$ is taken from Murphy [2013]. For reasons of simplicity, only one scenario exists such that the recourse function $Q(\chi)$ depends only on the first-stage variable χ:

$$\begin{aligned} Q(\chi) = &\min_{v} \ v \\ \text{s.t. } & v \geq \chi \\ & v \geq 10 - \chi \\ & v \geq 9 - 0.5\chi. \end{aligned}$$

This second-stage problem is illustrated in Figure A.1, where the three constraints (black lines) form the objective function $Q(\chi)$ (dashed line), which is piece-wise linear.

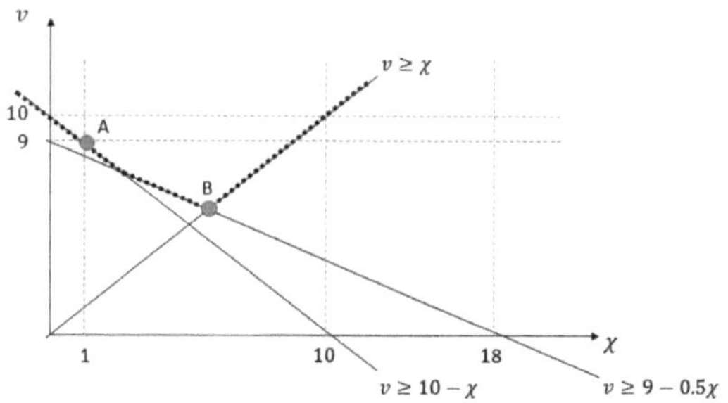

Figure A.1: Piece-wise linear and convex recourse function $Q(\chi)$ for the given second-stage problem

For the first-stage decision $\chi = 1$ the corresponding optimal second-stage objective function value is $v = 9$ (point A in Figure A.1). Obviously the minimum of this second-stage problem is at point B for $\chi = 6$, leading to a minimum second-stage objective function value of $v = 6$.

E. Graß, *An Accelerated Solution Method for Two-Stage Stochastic Models in Disaster Management*, Mathematische Optimierung und Wirtschaftsmathematik | Mathematical Optimization and Economathematics,
https://doi.org/10.1007/978-3-658-24081-3

A.2 Newton's Method for Systems of Non-Linear Equations

This description of Newton's method for solving systems of non-linear equations follows Vanderbei [2014, pp. 270].

Newton's method is among the most frequently used methods to find a root of a non-linear function, namely to find x in

$$\mathbf{f}(x) = 0,$$

with the vector-valued function $\mathbf{f} : \mathbb{R}^n \rightarrow \mathbb{R}^n$, i.e.

$$\mathbf{f}(x) = \begin{bmatrix} \mathbf{f}_1(x) \\ \mathbf{f}_2(x) \\ \vdots \\ \mathbf{f}_n(x) \end{bmatrix} = \begin{bmatrix} 0 \\ 0 \\ \vdots \\ 0 \end{bmatrix}, \text{ where } x = \begin{bmatrix} x_1 \\ x_2 \\ \vdots \\ x_n \end{bmatrix} \in \mathbb{R}^n.$$

The method improves the approximation x^k of a root x^* at each iteration k. The non-linear function $\mathbf{f}(x)$ is linearized via the use of the Taylor's series expansion, simplifying the root calculations. In particular, the Taylor series is terminated after the first two terms leading to linear approximations

$$\mathbf{f}(x^k + \triangle x^k) \approx \mathbf{f}(x^k) + \triangledown \mathbf{f}(x^k) \triangle x^k, \tag{A.1}$$

where

$$\triangledown \mathbf{f}(x^k) = \begin{bmatrix} \triangledown \mathbf{f}_1(x^k) \\ \triangledown \mathbf{f}_2(x^k) \\ \vdots \\ \triangledown \mathbf{f}_n(x^k) \end{bmatrix} = \begin{bmatrix} \frac{\partial \mathbf{f}_1}{\partial x_1} & \frac{\partial \mathbf{f}_1}{\partial x_2} & \cdots & \frac{\partial \mathbf{f}_1}{\partial x_n} \\ \frac{\partial \mathbf{f}_2}{\partial x_1} & \frac{\partial \mathbf{f}_2}{\partial x_2} & \cdots & \frac{\partial \mathbf{f}_2}{\partial x_n} \\ \vdots & \vdots & & \vdots \\ \frac{\partial \mathbf{f}_n}{\partial x_1} & \frac{\partial \mathbf{f}_n}{\partial x_2} & \cdots & \frac{\partial \mathbf{f}_n}{\partial x_n} \end{bmatrix}$$

is the *Jacobian matrix* of $\mathbf{f}$, containing all first-order partial derivatives. In Newton's method the search direction is determined by setting (A.1) to 0:

$$\mathbf{f}(x^k + \triangle x^k) \approx \mathbf{f}(x^k) + \triangledown \mathbf{f}(x^k) \triangle x^k = 0,$$

such that

$$\triangledown \mathbf{f}(x^k) \triangle x^k = -\mathbf{f}(x^k). \tag{A.2}$$

The Newton step is then

$$\triangle x^k = -\triangledown \mathbf{f}(x^k)^{-1} \mathbf{f}(x^k).$$

The new iterate x^{k+1} is set to:

$$x^{k+1} = x^k + \triangle x^k.$$

The algorithm terminates if a root is approximately found, i.e. $\mathbf{f}(x^k) \approx 0$. Applying Newton's method to the system (3.31)-(3.33) with

$$x = \begin{bmatrix} \gamma \\ \pi \\ z \end{bmatrix}, \ \mathbf{f}(x) = \begin{bmatrix} W^T\pi + z - f \\ W\gamma - \tilde{t} \\ \Gamma Ze - \mu e \end{bmatrix} \text{ and } \nabla \mathbf{f}(x) = \begin{bmatrix} 0 & W^T & I \\ W & 0 & 0 \\ Z & 0 & \Gamma \end{bmatrix}$$

leads, according to (A.2), to

$$\begin{bmatrix} 0 & W^T & I \\ W & 0 & 0 \\ Z & 0 & \Gamma \end{bmatrix} \begin{bmatrix} \triangle\gamma \\ \triangle\pi \\ \triangle z \end{bmatrix} = \begin{bmatrix} f - W^T\pi - z \\ \tilde{t} - W\gamma \\ -\Gamma Ze + \mu e \end{bmatrix},$$

see (3.35).

A.3 Interior-Point Method: Proof of Convergence

The proof of convergence follows Nocedal and Wright [2006, pp. 404]. Some definitions are required before giving the convergence proof. See Section 3.3.3 for notation.

Definition A.3.1. *The strictly feasible primal-dual set $\mathcal{F}$ is defined as:*

$$\mathcal{F} = \{(\gamma, \pi, z) | W\gamma = \tilde{t}, W^T\pi + z = f, (\gamma, z) > 0\}.$$

I.e. the point (γ, π, z) belongs to $\mathcal{F}$ if dual feasibility (3.31) and primal feasibility (3.32) are fulfilled and γ and z are strictly positive.

Definition A.3.2. *The $-\infty$-norm neighborhood $\mathcal{N}_{-\infty}(\varrho)$ of the central path $\mathcal{C}$ is defined as:*

$$\mathcal{N}_{-\infty}(\varrho) = \{(\gamma, \pi, z) \in \mathcal{F} | \Gamma Ze \geq \varrho\tau e\},$$

where $\varrho \in (0, 1]$.

Note that $\Gamma Ze \geq \varrho\tau e$ is equivalent to $\gamma_j z_j \geq \varrho\tau$ for $j = 1, ..., n_2$.

Lemma A.3.1. *If the step direction is determined by (3.35), then*[1]

$$\tau^{\mathbf{k+1}} = (1 - \alpha(1 - \sigma^{\mathbf{k}}))\tau^{\mathbf{k}}.$$

Proof. The last row of (3.35) is $Z\triangle\gamma + \Gamma\triangle z = -\Gamma Ze + \sigma\tau e$ with $\tau = \gamma^T z / n_2$ or equivalently in vector notation in iteration $\mathbf{k}$:

$$(z^{\mathbf{k}})^T \triangle\gamma^{\mathbf{k}} + (\gamma^{\mathbf{k}})^T \triangle z^{\mathbf{k}} = -(1 - \sigma^{\mathbf{k}})(\gamma^{\mathbf{k}})^T z^{\mathbf{k}}.$$

[1] Note that α represents a step length parameter which is not necessarily defined as in (3.37).

With this relation and the fact that $(\triangle\gamma^{\mathbf{k}})^T\triangle z^{\mathbf{k}} = 0$ [Wright, 1997, p. 87] it follows:

$$\begin{aligned}(\gamma^{\mathbf{k+1}})^T z^{\mathbf{k+1}} &= (\gamma^{\mathbf{k}} + \alpha\triangle\gamma^{\mathbf{k}})^T(z^{\mathbf{k}} + \alpha\triangle z^{\mathbf{k}})\\ &= (\gamma^{\mathbf{k}})^T z^{\mathbf{k}} + \alpha((\gamma^{\mathbf{k}})^T\triangle z^{\mathbf{k}} + (z^{\mathbf{k}})^T\triangle\gamma^{\mathbf{k}}) + \alpha^2(\triangle\gamma^{\mathbf{k}})^T\triangle z^{\mathbf{k}}\\ &= (\gamma^{\mathbf{k}})^T z^{\mathbf{k}} + \alpha(-(1-\sigma^{\mathbf{k}})(\gamma^{\mathbf{k}})^T z^{\mathbf{k}})\\ &= (\gamma^{\mathbf{k}})^T z^{\mathbf{k}}(1-\alpha(1-\sigma^{\mathbf{k}})).\end{aligned}$$

Dividing both sides by n_2 and $\tau^{\mathbf{k}} = \dfrac{(\gamma^{\mathbf{k}})^T z^{\mathbf{k}}}{n_2}$:

$$(\gamma^{\mathbf{k+1}})^T z^{\mathbf{k+1}}/n_2 = (\gamma^{\mathbf{k}})^T z^{\mathbf{k}}(1-\alpha(1-\sigma^{\mathbf{k}}))/n_2 \tag{A.3}$$

leads to

$$\tau^{\mathbf{k+1}} = (1-\alpha(1-\sigma^{\mathbf{k}}))\tau^{\mathbf{k}}.$$

□

Convergence of the path-following interior-point method in Figure 3.7 can be shown by proving that the duality measure $\tau^{\mathbf{k}}$ decreases in each iteration $\mathbf{k}$ such that $\tau^{\mathbf{k}} \to 0$ as $\mathbf{k} \to \infty$ [Nocedal and Wright, 2006, pp. 402].

Theorem A.3.1. *For a positive constant δ the following holds:*

$$\tau^{\mathbf{k+1}} \leq \left(1 - \frac{\delta}{n_2}\right)\tau^{\mathbf{k}} \quad \forall\mathbf{k} \geq 0 \tag{A.4}$$

Proof. First of all it has to be shown that the point $(\gamma^{\mathbf{k+1}}, \pi^{\mathbf{k+1}}, z^{\mathbf{k+1}})$ in the next iteration $\mathbf{k+1}$ is in the neighborhood of the central path $\mathcal{C}$, namely

$$(\gamma^{\mathbf{k+1}}, \pi^{\mathbf{k+1}}, z^{\mathbf{k+1}}) \in \mathcal{N}_{-\infty}(\varrho) \quad \forall\alpha \in \left[0,\ 2^{3/2}\varrho\frac{1-\varrho\,\sigma^{\mathbf{k}}}{1+\varrho\,n_2}\right].$$

For this purpose, (3.38) can be used to write the product $\gamma_j^{\mathbf{k+1}} z_j^{\mathbf{k+1}}$ for $j = 1, ..., n$ as:

$$\begin{aligned}\gamma_j^{\mathbf{k+1}} z_j^{\mathbf{k+1}} &= (\gamma_j^{\mathbf{k}} + \alpha\triangle\gamma_j^{\mathbf{k}})(z_j^{\mathbf{k}} + \alpha\triangle z_j^{\mathbf{k}})\\ &= \gamma_j^{\mathbf{k}} z_j^{\mathbf{k}} + \alpha(\gamma_j^{\mathbf{k}}\triangle z_j^{\mathbf{k}} + z_j^{\mathbf{k}}\triangle\gamma_j^{\mathbf{k}}) + \alpha^2\triangle\gamma_j^{\mathbf{k}}\triangle z_j^{\mathbf{k}}.\end{aligned}$$

From the last row of (3.35), i.e. $Z\triangle\gamma + \Gamma\triangle z = -\Gamma Ze + \sigma\tau e$, it follows:

$$\begin{aligned}\gamma_j^{\mathbf{k+1}} z_j^{\mathbf{k+1}} &= \gamma_j^{\mathbf{k}} z_j^{\mathbf{k}}(1-\alpha) + \alpha\sigma^{\mathbf{k}}\tau^{\mathbf{k}} + \alpha^2\triangle\gamma_j^{\mathbf{k}}\triangle z_j^{\mathbf{k}}\\ &\geq \gamma_j^{\mathbf{k}} z_j^{\mathbf{k}}(1-\alpha) + \alpha\sigma^{\mathbf{k}}\tau^{\mathbf{k}} - \alpha^2|\triangle\gamma_j^{\mathbf{k}}\triangle z_j^{\mathbf{k}}|\end{aligned} \tag{A.5}$$

The last line is used in order to apply the inequality given in Nocedal and Wright [2006, Lemma 14.2]:

$$|\triangle\gamma_j^{\mathbf{k}}\triangle z_j^{\mathbf{k}}| \leq \|\triangle\Gamma^{\mathbf{k}}\triangle Z^{\mathbf{k}}e\|_2 \leq 2^{-3/2}(1+1/\varrho)n_2\tau^{\mathbf{k}}.$$

In addition, according to definition A.3.2, i.e. $\gamma_j^{\mathbf{k}} z_j^{\mathbf{k}} \geq \varrho\tau^{\mathbf{k}}$, for $j = 1, ..., n_2$, (A.5) can be rewritten

$$\gamma_j^{\mathbf{k+1}} z_j^{\mathbf{k+1}} \geq \varrho\tau^{\mathbf{k}}(1-\alpha) + \alpha\sigma^{\mathbf{k}}\tau^{\mathbf{k}} - \alpha^2 2^{-3/2}(1+1/\varrho)n_2\tau^{\mathbf{k}}. \tag{A.6}$$

For

$$\alpha \leq \frac{2^{3/2}}{n_2}\sigma^{\mathbf{k}}\varrho\frac{1-\varrho}{1+\varrho}, \tag{A.7}$$

the following inequality holds

$$\varrho\tau^{\mathbf{k}}(1-\alpha) + \alpha\sigma^{\mathbf{k}}\tau^{\mathbf{k}} - \alpha^2 2^{-3/2}(1+1/\varrho)n_2\tau^{\mathbf{k}} \geq \varrho\tau^{\mathbf{k}}(1-\alpha+\alpha\sigma^{\mathbf{k}}). \tag{A.8}$$

In this case, using (A.8) and Lemma A.3.1, (A.6) can be formulated as follows:

$$\gamma_j^{\mathbf{k+1}} z_j^{\mathbf{k+1}} \geq \varrho\tau^{\mathbf{k+1}}.$$

As a result the next iteration $(\gamma^{\mathbf{k+1}}, \pi^{\mathbf{k+1}}, z^{\mathbf{k+1}})$ lies in the neighborhood $\mathcal{N}_{-\infty}(\varrho)$ for a step length α given in (A.7).

The next step is to show that the inequality (A.4) holds. For the step length $\alpha \geq \frac{2^{3/2}}{n_2}\sigma^{\mathbf{k}}\varrho\frac{1-\varrho}{1+\varrho}$ and the relation in Lemma A.3.1, $\tau^{\mathbf{k+1}}$ is as follows:

$$\tau^{\mathbf{k+1}} \leq (1 - \frac{2^{3/2}}{n_2}\varrho\frac{1-\varrho}{1+\varrho}\sigma^{\mathbf{k}}(1-\sigma^{\mathbf{k}}))\tau^{\mathbf{k}}. \tag{A.9}$$

By defining

$$\delta = 2^{3/2}\varrho\frac{1-\varrho}{1+\varrho}\sigma^{\mathbf{k}}(1-\sigma^{\mathbf{k}}),$$

where $\sigma^{\mathbf{k}}(1-\sigma^{\mathbf{k}}) \geq \min\{\sigma_{min}(1-\sigma_{min}), \sigma_{max}(1-\sigma_{max})\}$, $\forall\sigma^{\mathbf{k}} \in [\sigma_{min}, \sigma_{max}]$ [Nocedal and Wright, 2006, p. 405], (A.9) is

$$\tau^{\mathbf{k+1}} \leq (1 - \frac{\delta}{n_2})\tau^{\mathbf{k}},$$

which is equivalent to (A.4). □

Finally, it is necessary to show that the reduction of $\tau^{\mathbf{k}}$ is finite requiring $O(n \log 1/\varepsilon)$ iterations. Therefore, the interior-point method described in Figure 3.7 has the following polynomial complexity property.

Theorem A.3.2. *For* $\varepsilon \in (0,1)$, $\varrho \in (0,1)$ *and* $(\gamma^0, \pi^0, z^0) \in \mathcal{N}_{-\infty}(\varrho)$ *there exists an iteration number* $\mathbf{K} = O(n \log 1/\varepsilon)$ *such that*

$$\tau^{\mathbf{k}} \leq \varepsilon\tau^0 \quad \forall\mathbf{k} \geq \mathbf{K}.$$

Proof. Using the logarithm, (A.4) becomes

$$\log \tau^{\mathbf{k+1}} \leq \log\left(1-\frac{\delta}{n_2}\right)\tau^{\mathbf{k}} = \log\left(1-\frac{\delta}{n_2}\right) + \log \tau^{\mathbf{k}}.$$

Repeating the formula recursively for $\log \tau^{\mathbf{k}}$, it follows

$$\log \tau^{\mathbf{k}} \leq \mathbf{k} \log\left(1-\frac{\delta}{n_2}\right) + \log \tau^0,$$

or equivalently

$$\log \tau^{\mathbf{k}} - \log \tau^0 \leq \mathbf{k} \log\left(1-\frac{\delta}{n_2}\right).$$

The logarithm rules $\log x - \log y = \log(x/y)$ and $\log(1+\beta) \leq \beta$ for $\beta > -1$ lead to

$$\log\left(\frac{\tau^{\mathbf{k}}}{\tau^0}\right) \leq \mathbf{k}\left(-\frac{\delta}{n_2}\right).$$

Since it has to be shown that $\tau^{\mathbf{k}}/\tau^0 \leq \epsilon$ or equivalently $\log(\tau^{\mathbf{k}}/\tau^0) \leq \log \epsilon$, the following should hold:

$$\mathbf{k}\left(-\frac{\delta}{n_2}\right) \leq \log \epsilon.$$

This is true if $\mathbf{k} \geq \mathbf{K}$, where

$$\mathbf{K} = \frac{n_2}{\delta} \log \frac{1}{\epsilon} = \frac{n_2}{\delta} |\log \epsilon|.$$

□

A.4 Matlab Code: L-Shaped Method with Multi-Optimality Cuts

Function: LShaped_MultiCut.m

```
clear all
clc
% L-shaped method with multi-optimality cuts
%% Initial settings
% Choose test problem
testprob=2; % 1= Small case study based on RT (2010);
            % 2= Medium case study;
            % 3= Large case study
            % 4= Katrina case study
```

```

% Choose maximal number of L-shaped iterations
n=30;

% Choose tolerance level for the L-shaped method
epsilon=1e-04;

% Choose tolerance level for the master problem
GAP=1e-04;

% Choose solution method for sub-problems
sol=1; %1 = Specialized Interior-Point Method SIPM; 2 = built
    -in function "linprog"

% Choose maximal number of iterations for SIPM
 maxit_IP=100;

options = optimoptions('linprog','Algorithm','interior-point-
    legacy', 'OptimalityTolerance', 1e-05);

%% Range of test problems
% Small case study based on RT (2010)
if testprob == 1
nodes=30;               % Set of nodes: N
scenarios=51;           % Set of scenarios: S
commodities=3;          % Set of relief Iiems: K
locTypes = 3;           % Set of facility size categories: L
arcs=nodes*nodes;       % Set of arcs: R

load('RT_CaseStudy.mat')
prob=full(prob);
a=a';

% Medium-scale case study
elseif testprob == 2
nodes = 90;             % Set of nodes: N
locTypes = 3;           % Set of facility size categories: L
commodities =3;         % Set of relief Iiems: K
scenarios = 80;         % Set of scenarios: S
arcs=nodes*nodes;       % Set of arcs: R

load('MediumCaseStudy.mat')
prob=full(prob);
a=a';
```

```

% Large-scale case study
elseif testprob == 3
nodes=90;              % Set of nodes: N
scenarios=118;         % Set of scenarios: S
commodities=5;         % Set of relief Iiems: K
locTypes = 3;          % Set of facility size categories: L
arcs=nodes*nodes;      % Set of arcs: R

load('LargeCaseStudy.mat')
prob=full(prob);
a=a';

% Katrina case study
elseif testprob == 4
nodes=90;              % Set of nodes: N
scenarios=118;         % Set of scenarios: S
commodities=5;         % Set of relief Iiems: K
locTypes = 3;          % Set of facility size categories: L
arcs=nodes*nodes;      % Set of arcs: R

load('ShortTermCaseStudy.mat')
prob=full(prob);
a=a';
end

% Preallocation for speed
t_s_new=zeros(arcs+nodes*commodities,1,scenarios);
gamma=zeros(arcs*commodities+2*nodes*commodities+arcs,1,
    scenarios);
pi=zeros(arcs+nodes*commodities,1,scenarios);
objval=zeros(1,1,scenarios);
lambda=zeros(1,scenarios);
e=zeros(scenarios,1);
E=zeros(scenarios,nodes*(commodities+locTypes));
ev=zeros(1,1,scenarios);
Ev=zeros(1,nodes*(commodities+locTypes),scenarios);
gammav=zeros(1,1,scenarios);

%% L-shaped method
%% Step 0: Initialization
k=0;
v=1;           % L-shaped iteration counter
UB(v)=1e+16;   % Initial upper bound
```

```
LB(v)=-UB(v); % Initial lower bound

lb=zeros(nodes*(locTypes+commodities),1);  % First stage: non negativity
fnrm=norm(f);                              % Used for initial point generation

% Main loop
% Stop if RelOptGap < epsilon or maximal number of iterations reached
while (UB(v)-LB(v))>epsilon*UB(v) && v<n
RelOptGap=(UB(v)-LB(v))/UB(v) % show current RelOptGap

%% Step 1: Solving master problem
intcon=1:nodes*locTypes;  % binary variables

% Gurobi solver with output x: solution vector; fval: objective value
[x,fval] = intlinprog_gurobi(full(a'),intcon,G,full(d),[],[],lb,[],GAP);

% L-shaped iteration count update
v=v+1

% LB update
LB(v)=fval;

%% Step 2: Solving sub-problem for every scenario s
x=x(1:nodes*(locTypes+commodities));

for s=1:scenarios
    s

% Right hand side
t_s_new(:,:,s)=t((s-1)*(arcs+nodes*commodities)+1:s*(arcs+nodes*commodities))-T((s-1)*(arcs+nodes*commodities)+1:s*(arcs+nodes*commodities),:)*x;

%% Solution Method
 if sol==1 % SIPM

 tnrm=norm(t_s_new(:,:,s));

 % Generating initial point (Zhang, 1998)
```

```
[gamma0, pi0 , z0]=initpointpos(W, t_s_new (: ,: , s) ,f ,tnrm, fnrm) ;

  [gamma(: ,: , s) ,  pi (: ,: , s) ,  objval (: ,: , s) , iter , . . .
        normprimal ,  normdual ,  optimalitygap ] = . . .
   SIPM(f , W, t_s_new (: ,: , s) , u, H, Q, nodes , arcs ,
       commodities ,gamma0 , . . .
        pi0 , z0 , maxit_IP ) ;


elseif sol==2 % MATLAB's built-in function "linprog"
% linprog
% min c'T x
% s.t. Ax<=b
%      Aeq*x=beq
%      lb<=x<=ub
% syntax: x=linprog(c,A,b,Aeq,beq,lb,ub)

% Non-negativity of second-stage variables
lbsec=sparse(2*nodes*commodities+arcs*commodities+arcs ,1) ;
[gamma(: ,: , s) ,objval (: ,: , s) ,~ , ~ , lambda]=linprog(f ,[] ,[] ,W
    , . . .
      t_s_new (: ,: , s) ,lbsec ,[] ,[] , options) ;

pi (: ,: , s)=-lambda.eqlin ; % Simplex multiplier
end

%% Step 3: Generating optimality cuts
ev (: ,: , s)=prob(s)*(pi (: ,: , s) '*t((s-1)*(arcs+nodes*commodities
    )+1:s*(arcs+nodes*commodities))) ;
Ev (: ,: , s)=prob(s)*(pi (: ,: , s) '*T((s-1)*(arcs+nodes*commodities
    )+1:s*(arcs+nodes*commodities) ,:)) ;
gammav (: ,: , s)=prob(s)*(objval (: ,: , s)) ;

end

for i=1:s
e(i ,1)=ev (: ,: , i ) ;
E(i ,:)=Ev (: ,: , i ) ;
end

w=sum(gammav) ; % Probability weighted second-stage objective
    function

% Upper bound update
```

```
UB(v)=min(UB(v-1),a(1:nodes*(locTypes+commodities))'*x+w);

k=k+1;

 if k==1
      G(:,end+1:end+s)=0;
      a(end+1:end+s)=1;
      lb(end+1:end+s)=-inf;
end

% rhs with added optimaltiy cuts (-e to convert >= into <= inequality)
d=vertcat(d,-e);
% constraint matrix G with added optimaltiy cuts (-E to convert >= into <=
% inequality)
G=vertcat(G,[-E -speye(s,s)]);

end
          fprintf ('final lower bound value (LB) %f \n', fval);
          fprintf ('final upper bound value (UB) %f \n', UB(v));
          fprintf ('number of L-Shaped iterations %f \n', v);
```

Function: initpointpos.m

```
function [x,y,z] = initpointpos(A,b,c,bnrm,cnrm)
% Specifying the starting point
% Yin Zhang, January, 1995
% Department of Mathematics and Statistics
% University of Maryland  Baltimore County

[m,n] = size(A); y = sparse(m,1);
pmin = max(bnrm/100, 100);
dmin = cnrm*.425; dmin = max(dmin, pmin/40);
pmin = min(pmin, 1.e+4); dmin = min(dmin, 1.e+3);

e = ones(n,1);
    rho = min(100,bnrm);
    x = A'*(b-rho*A*e) + rho*e;
pmin = max(pmin, -min(x)); x = max(pmin,x);
z = (c+dmin).*(c > 0) + dmin*(-dmin < c & c <= 0) - c.*(c <= -dmin);
```

A.5 Matlab Code: SIMP

Function: SIPM.m

```
function [gamma_sol, pi, objval, iter, ...
          normprimal, normdual, optimalitygap] = ...
SIPM(f, W, t_new, u, H, Q, nodes, arcs, commodities,gamma0, pi0,z0,maxit_IP)
% SIPM solves a linear programming in standard form
% Primal: min f'*gamma               Dual: max t_new'pi
%         s.t. W*gamma=t_new               s.t. W'*pi+z=f
%              gamma >=0                        z >=0
% using the primal-dual infeasible path following method
%
% Main Inputs:
%  f = vector of objective coefficients
%  W = Recourse matrix with special lower block triangular structure
%  t_new = RHS-vector
%  gamma0 = inital vector for gamma
%  pi0 = inital vector for pi
%  z0 = inital vector for z
%  maxit_IP = maximal number of iterations for SIPM
%
% Main Outputs:
%  gamm_sol = final solution vector
%  pi= = vector of Simplex multiplier
%  objval = final objective value
%  iter = required number of iterations

%% Step 0: Initialization
%%%%%%%%%%%%%%%%%%%%%%%%%%%%%%%%
% Obtain an initial interior solution [gamma(0), pi(0), z(0)]^T such that
% gamma(0)>0 and z(0)>0. Let k=0

% Step length dampening constant (Ferris et al., 2007)
etaMin=.9995;

% Default tolerance level for relative optimality gap
epsi_opt=1e-5;

% Flag for outer loops; as long as matrix A is not ill-conditioned
ill=false;
```

```
[~, n]=size(W); % Number of variables

% Interior starting point
gamma=gamma0;    % Initial gamma(0), vector of primal variables
pi=pi0;          % Initial pi(0), vector of dual variables / Simplex multiplier
z=z0;            % Initial z(0), vector of slack variables of the dual problem

for iter=1:maxit_IP
%% Step 1: Generating search direction

mu_average=(gamma'*z)/(n); % duality measure

% Relative primal residuals
normprimal(iter) = norm(W*gamma-t_new)/(max(1,norm(t_new)));

% Relative dual residuals
normdual(iter) = norm(W'*pi+z-f)/(max(1,norm(f)));

% Relative optimality gap
optimalitygap(iter) = abs(f'*gamma-(t_new'*pi))/(max(1,abs(f'*gamma)));

% Stop if sum of relative residuals and optimality gap <= epsi_opt or
% system becomes ill-conditioned or maximal number of iterations reached
 if (normprimal(iter)+ normdual(iter)+optimalitygap(iter) <= epsi_opt) || ill
     disp('problem solved')
 break
 elseif iter>=maxit_IP
    disp('maximal number of iterations is reached')
 break
 end;

 % centering parameter (Ferris et al., 2007)
 sigma=min(0.1,100*mu_average);

 % barrier parameter mu
 mu=sigma*mu_average;
```

```
 % Right hand sides
 rhs_d=f-W'*pi-mu./gamma;                    % rhs: dual residuals
 rhs_p=t_new-W*gamma;                        % rhs: primal residuals

%% Generating block matrices of WD^-1W^T
zg=z.\gamma; % Inverse of diagonal entries z^-1*gamma, i.e. D^-1

% (1,1)-block A
Du=zg(1:end-2*nodes*commodities).* repmat(u',arcs,1);
Du_re=reshape(Du,[commodities+1,arcs]);
A=(u*Du_re)'; % stored as a vector

% (1,2)-block C
ii=1:numel(Du_re);
jj=repmat(1:arcs, commodities+1,1);
Du_diag=sparse(ii',jj(:),Du_re(:));
C=H*Du_diag;

% (2,2)-block B
HQ=[H Q];
B=HQ*(spdiags(zg,0,n,n)*HQ');

% rhs update
rhs_new=rhs_p+W*(zg.*rhs_d);

Amat=spdiags(A,0,arcs,arcs);
AC=Amat\C';

 % Checking for ill-conditioning
  [~,msgid]=lastwarn;
   if strcmp(msgid,'MATLAB:nearlySingularMatrix')
  ill=true;
  end

rhs_1=rhs_new(1:arcs);
rhs_2=rhs_new(arcs+1:end);

% Schur complement of WD^-1W^T
Schur=B-(C*AC);
```

```
% Using Cholesky factorization to obtain an upper triangular matrix,
% such that UT'*UT=Schur
UT=chol(Schur);

% Solving system Schur*x=b via forward and
% backward substitution: UT'*y=b and UT*x=y
d_pi_2=UT\(UT'\(rhs_2-C*(Amat\rhs_1))); %

%% Recover the other directions
d_pi_1=A.\(rhs_1-C'*d_pi_2);

d_pi=[d_pi_1;d_pi_2];                    % direction of pi
d_gamma=(-zg).*(rhs_d-W'*d_pi);          % direction of gamma
d_z=mu./gamma-z-(gamma.\z).*d_gamma;     % direction of z

lastwarn('')
%%%%%%%%%%%%%%%%%%%%%%%%%%%%%%%%%%%%%%%%%%%%%%%%
%% Step 2: Update Solution
%%%%%%%%%%%%%%%%%%%%%%%%%%%%%%%%%%%%%%%%%%%%%%%%

% Step length dampening parameter (Ferris et al., 2007)
  eta = max(etaMin,1-mu_average);

% Ratio test (Feris et al., 2007)
  [alphap,alphad] = steplength(gamma,z,d_gamma,d_z, eta);

% Solution update
  gamma = gamma + alphap * d_gamma;
  z = z + alphad * d_z;
  pi = pi + alphad * d_pi;

end
gamma_sol=gamma;
objval=f'*gamma;
```

Function: steplength.m

```
function [alphap, alphad] = steplength(x,z,dx,dz,eta)
% Compute steplengths ensuring x + alphap*d_x>0 and z + alphad*d_z>0

alphap = -1/min(min(dx./x),-1); alphap = min(1, eta * alphap);
```

```
alphad = -1/min(min(dz./z),-1); alphad = min(1, eta * alphad)
    ;
```

A.6 Gurobi Log Files

A.6.1 Small-Scale Case Study

```
Gurobi 7.0.0 (win64) logging started 07/19/17 11:06:14

Set parameter Method to value 2

Gurobi Optimizer version 7.0.0 build v7.0.0rc4 (win64)
Copyright (c) 2016, Gurobi Optimization, Inc.

Read MPS format model from file RT2010MIP.mps
Reading time = 0.45 seconds
RT2010MI: 50550 rows, 147060 columns, 417777 nonzeros
Optimize a model with 50550 rows, 147060 columns and 417777 nonzeros
Variable types: 146970 continuous, 90 integer (0 binary)
Coefficient statistics:
  Matrix range     [5e-01, 8e+05]
  Objective range  [1e-04, 3e+05]
  Bounds range     [1e+00, 1e+00]
  RHS range        [3e-01, 1e+06]
Found heuristic solution: objective 1.68213e+08
Presolve removed 40457 rows and 123863 columns
Presolve time: 0.27s
Presolved: 10093 rows, 23197 columns, 60413 nonzeros
Variable types: 23122 continuous, 75 integer (75 binary)
Presolve removed 4 rows and 4 columns
Presolved: 10089 rows, 23193 columns, 60397 nonzeros

Root barrier log...

Ordering time: 0.17s

Barrier statistics:
 AA' NZ     : 1.506e+05
 Factor NZ  : 1.717e+06 (roughly 30 MBytes of memory)
 Factor Ops : 7.360e+08 (less than 1 second per iteration)
 Threads    : 2

                  Objective                Residual
Iter       Primal          Dual         Primal    Dual     Compl     Time
   0   5.31905895e+12 -2.12093430e+14  1.18e+06 2.78e+03  6.75e+09     1s
```

```
   1   2.96578215e+12 -5.50583796e+13  6.45e+05 2.39e+03  2.21e+09     1s
   2   1.03254338e+12 -1.16251317e+13  2.00e+05 4.36e+02  5.73e+08     1s
   3   2.24458980e+11 -3.24645959e+11  2.07e+04 2.33e-10  4.26e+07     1s
   4   4.96820791e+10 -6.85895701e+10  2.23e+03 5.25e-01  5.54e+06     1s
   5   6.13085049e+09 -1.37075057e+10  1.82e+02 1.46e-10  6.27e+05     1s
   6   3.15900008e+09 -9.34816086e+09  8.14e+01 8.73e-11  3.41e+05     1s
   7   5.90716774e+08 -2.24495189e+09  3.39e+00 1.16e-10  5.58e+04     1s
   8   3.80477818e+08 -1.15357142e+09  1.57e+00 5.82e-11  2.92e+04     1s
   9   2.99248003e+08 -8.37294767e+08  9.96e-01 5.82e-11  2.14e+04     1s
  10   2.45351934e+08 -6.99700127e+08  5.99e-01 5.82e-11  1.76e+04     1s
  11   2.24808011e+08 -4.87307492e+08  4.63e-01 5.82e-11  1.32e+04     1s
  12   2.00791938e+08 -2.40113920e+08  3.29e-01 5.82e-11  8.17e+03     1s
  13   1.64434059e+08 -1.13480053e+08  1.28e-01 5.82e-11  5.12e+03     1s
  14   1.45654205e+08  3.60430121e+07  5.38e-02 5.82e-11  2.02e+03     1s
  15   1.35870108e+08  9.31840301e+07  2.22e-02 5.82e-11  7.85e+02     1s
  16   1.30450207e+08  1.05214587e+08  1.06e-02 6.18e-11  4.64e+02     1s
  17   1.28633627e+08  1.13752253e+08  6.74e-03 5.82e-11  2.73e+02     1s
  18   1.27344412e+08  1.20525126e+08  3.62e-03 6.55e-11  1.25e+02     1s
  19   1.26468559e+08  1.24117878e+08  1.33e-03 5.37e-11  4.32e+01     2s
  20   1.26088203e+08  1.25157318e+08  4.41e-04 5.82e-11  1.71e+01     2s
  21   1.25967205e+08  1.25596164e+08  1.86e-04 6.72e-11  6.82e+00     2s
  22   1.25908503e+08  1.25736699e+08  7.33e-05 6.22e-11  3.16e+00     2s
  23   1.25892990e+08  1.25794973e+08  4.63e-05 6.82e-11  1.80e+00     2s
  24   1.25878267e+08  1.25830928e+08  2.31e-05 5.98e-11  8.70e-01     2s
  25   1.25865274e+08  1.25851047e+08  5.12e-06 1.07e-10  2.61e-01     2s
  26   1.25860882e+08  1.25857428e+08  7.12e-07 1.21e-10  6.34e-02     2s
  27   1.25860083e+08  1.25858877e+08  2.75e-07 2.91e-11  2.22e-02     2s
  28   1.25859642e+08  1.25859382e+08  4.89e-08 1.16e-10  4.78e-03     2s
  29   1.25859556e+08  1.25859470e+08  5.26e-08 7.05e-11  1.57e-03     2s
  30   1.25859530e+08  1.25859495e+08  2.09e-08 8.00e-11  6.53e-04     2s
  31   1.25859516e+08  1.25859504e+08  5.01e-09 5.82e-11  2.33e-04     2s
  32   1.25859512e+08  1.25859511e+08  6.90e-08 5.82e-11  2.61e-05     2s
  33   1.25859512e+08  1.25859512e+08  7.29e-07 5.74e-11  1.91e-06     2s
  34   1.25859512e+08  1.25859512e+08  1.97e-08 1.16e-10  5.65e-09     2s

Barrier solved model in 34 iterations and 2.29 seconds
Optimal objective 1.25859512e+08


Root relaxation: objective 1.258595e+08, 5160 iterations, 2.03 seconds

    Nodes    |    Current Node    |     Objective Bounds      |     Work
 Expl Unexpl |  Obj  Depth IntInf | Incumbent    BestBd   Gap | It/Node Time

     0     0 1.2586e+08    0   23 1.6821e+08 1.2586e+08  25.2%     -    2s
H    0     0                    1.273985e+08 1.2586e+08  1.21%     -    2s
     0     0 1.2600e+08    0   27 1.2740e+08 1.2600e+08  1.10%     -    2s
```

```
     0     0 1.2613e+08    0   25 1.2740e+08 1.2613e+08  1.00%     -    3s
     0     0 1.2626e+08    0   27 1.2740e+08 1.2626e+08  0.89%     -    3s
H    0     0                    1.268085e+08 1.2626e+08  0.43%     -    3s
     0     0 1.2629e+08    0   22 1.2681e+08 1.2629e+08  0.41%     -    3s
     0     0 1.2633e+08    0   18 1.2681e+08 1.2633e+08  0.37%     -    3s
     0     0 1.2634e+08    0   18 1.2681e+08 1.2634e+08  0.37%     -    3s
     0     0 1.2635e+08    0   18 1.2681e+08 1.2635e+08  0.36%     -    3s
     0     0 1.2636e+08    0   17 1.2681e+08 1.2636e+08  0.35%     -    3s
     0     0 1.2636e+08    0   17 1.2681e+08 1.2636e+08  0.35%     -    3s
     0     0 1.2636e+08    0   15 1.2681e+08 1.2636e+08  0.35%     -    3s
     0     0 1.2636e+08    0   17 1.2681e+08 1.2636e+08  0.35%     -    3s
     0     0 1.2637e+08    0   17 1.2681e+08 1.2637e+08  0.35%     -    4s
     0     0 1.2637e+08    0   17 1.2681e+08 1.2637e+08  0.35%     -    4s
     0     0 1.2637e+08    0   15 1.2681e+08 1.2637e+08  0.35%     -    4s
     0     0 1.2637e+08    0   15 1.2681e+08 1.2637e+08  0.35%     -    4s
     0     0 1.2637e+08    0   15 1.2681e+08 1.2637e+08  0.35%     -    4s
     0     0 1.2637e+08    0   15 1.2681e+08 1.2637e+08  0.35%     -    4s
     0     0 1.2637e+08    0   15 1.2681e+08 1.2637e+08  0.35%     -    4s
     0     0 1.2637e+08    0   15 1.2681e+08 1.2637e+08  0.35%     -    4s
     0     2 1.2637e+08    0   15 1.2681e+08 1.2637e+08  0.35%     -    4s
     9    12 1.2652e+08    4   10 1.2681e+08 1.2640e+08  0.32%   349    5s
H   32    31                    1.267281e+08 1.2640e+08  0.26%   274    5s
H   45    24                    1.265530e+08 1.2640e+08  0.12%   262    5s
H   52    18                    1.265156e+08 1.2641e+08  0.08%   312    6s
H   57    18                    1.265007e+08 1.2641e+08  0.07%   317    6s
*   61    20              27    1.265005e+08 1.2641e+08  0.07%   322    6s
H   76    26                    1.264995e+08 1.2643e+08  0.05%   318    6s
H   83    21                    1.264705e+08 1.2643e+08  0.03%   311    6s
H   84    21                    1.264691e+08 1.2643e+08  0.03%   312    6s
*   85    20              21    1.264673e+08 1.2643e+08  0.03%   [illegible]    6s
*  105    17              19    1.264649e+08 1.2643e+08  0.02%   313    7s
*  110    14              16    1.264599e+08 1.2643e+08  0.02%   308    7s

Cutting planes:
  Gomory: 14
  Cover: 3
  Implied bound: 61
  MIR: 48
  Flow cover: 2

Explored 161 nodes (57279 simplex iterations) in 7.78 seconds
Thread count was 4 (of 4 available processors)

Solution count 10: 1.2646e+08 1.26465e+08 1.26467e+08 ... 1.26553e+08
Pool objective bound 1.26448e+08

Optimal solution found (tolerance 1.00e-04)
```

```
Best objective 1.264599354391e+08, best bound 1.264475821813e+08, gap 0.0098%

Wrote result file 'SmallAdjSol.sol'
```

A.6.2 Medium-Scale Case Study

```
Gurobi 7.0.0 (win64) logging started 07/19/17 09:40:48

Set parameter MIPGap to value 0.07
Set parameter Method to value 2

Gurobi Optimizer version 7.0.0 build v7.0.0rc4 (win64)
Copyright (c) 2016, Gurobi Optimization, Inc.

Read LP format model from file Medium.lp
Reading time = 13.92 seconds
P1: 669780 rows, 1987740 columns, 5854282 nonzeros
Optimize a model with 669780 rows, 1987740 columns and 5854282 nonzeros
Variable types: 1987470 continuous, 270 integer (0 binary)
Coefficient statistics:
  Matrix range     [3e-03, 8e+05]
  Objective range  [1e-03, 3e+05]
  Bounds range     [1e+00, 1e+00]
  RHS range        [3e-01, 7e+06]
Found heuristic solution: objective 1.85209e+08
Presolve removed 14087 rows and 314559 columns (presolve time = 5s) ...
Presolve removed 14144 rows and 314559 columns (presolve time = 11s) ...
Presolve removed 14144 rows and 314559 columns
Presolve time: 11.11s
Presolved: 655636 rows, 1673181 columns, 4954024 nonzeros
Variable types: 1672911 continuous, 270 integer (270 binary)
Presolved: 655636 rows, 1673181 columns, 4954024 nonzeros

Root barrier log...

Elapsed ordering time = 9s
Elapsed ordering time = 12s
Elapsed ordering time = 15s
Elapsed ordering time = 24s
Elapsed ordering time = 26s
Elapsed ordering time = 30s
Elapsed ordering time = 40s
Elapsed ordering time = 41s
Elapsed ordering time = 45s
Elapsed ordering time = 54s
```

```
Elapsed ordering time = 56s
Elapsed ordering time = 60s
Elapsed ordering time = 69s
Elapsed ordering time = 71s
Elapsed ordering time = 75s
Ordering time: 77.05s

Barrier statistics:
 AA' NZ     : 4.938e+06
 Factor NZ  : 1.687e+08 (roughly 2.3 GBytes of memory)
 Factor Ops : 1.620e+12 (roughly 26 seconds per iteration)
 Threads    : 2

                  Objective                Residual
Iter       Primal          Dual         Primal    Dual     Compl     Time
   0   1.82612287e+15 -1.22111631e+16  1.50e+08 2.19e+01  1.11e+10   129s
   1   4.49714866e+13 -2.44939879e+14  3.37e+06 1.16e-10  1.76e+08   170s
   2   5.49946040e+12 -4.62735747e+12  3.35e+05 5.82e-11  1.26e+07   210s
   3   2.19140487e+11 -4.95626106e+11  3.20e+03 1.16e-10  2.76e+05   252s
   4   3.30806875e+10 -1.52188655e+11  1.06e+02 1.16e-10  4.77e+04   295s
   5   7.27190264e+09 -1.00953996e+10  1.28e+01 1.16e-10  4.37e+03   334s
   6   4.21614331e+09 -3.91065432e+09  5.63e+00 5.82e-11  2.04e+03   373s
   7   2.17611270e+09 -9.08593108e+08  2.11e+00 5.82e-11  7.75e+02   414s
   8   1.56837917e+09 -4.96052732e+08  1.38e+00 5.82e-11  5.19e+02   453s
   9   1.03694158e+09 -1.95834798e+08  8.29e-01 5.82e-11  3.10e+02   495s
  10   8.28212344e+08 -7.40265698e+07  6.19e-01 5.82e-11  2.27e+02   531s
  11   7.83744979e+08 -2.83513215e+07  5.76e-01 5.82e-11  2.04e+02   569s
  12   6.40170860e+08  2.32830882e+07  4.48e-01 5.82e-11  1.55e+02   608s
  13   5.89223104e+08  4.43342073e+07  4.03e-01 5.82e-11  1.37e+02   646s
  14   4.92176321e+08  7.55684178e+07  3.10e-01 5.82e-11  1.05e+02   687s
  15   4.17175375e+08  8.25655906e+07  2.49e-01 5.82e-11  8.41e+01   724s
  16   3.56114344e+08  1.01238309e+08  1.95e-01 5.82e-11  6.40e+01   763s
  17   3.04913441e+08  1.10995201e+08  1.48e-01 5.82e-11  4.87e+01   801s
  18   2.47582080e+08  1.23429949e+08  9.34e-02 5.82e-11  3.12e+01   842s
  19   2.25809134e+08  1.26436146e+08  7.30e-02 5.82e-11  2.50e+01   881s
  20   2.04672408e+08  1.33828874e+08  5.26e-02 5.82e-11  1.78e+01   919s
  21   1.90585567e+08  1.40009083e+08  3.79e-02 5.82e-11  1.27e+01   957s
  22   1.77317069e+08  1.42027152e+08  2.59e-02 1.16e-10  8.87e+00   997s
  23   1.71520650e+08  1.44621337e+08  2.02e-02 5.82e-11  6.76e+00  1037s
  24   1.66568723e+08  1.45823509e+08  1.52e-02 5.82e-11  5.21e+00  1075s
  25   1.63606309e+08  1.46978857e+08  1.23e-02 5.82e-11  4.18e+00  1113s
  26   1.60522539e+08  1.47845647e+08  9.19e-03 5.82e-11  3.18e+00  1153s
  27   1.59701073e+08  1.48278254e+08  8.39e-03 5.82e-11  2.87e+00  1191s
  28   1.58577317e+08  1.48715458e+08  7.28e-03 5.82e-11  2.48e+00  1231s
  29   1.57124002e+08  1.49201757e+08  5.81e-03 5.82e-11  1.99e+00  1271s
  30   1.56200010e+08  1.49477555e+08  4.88e-03 1.16e-10  1.69e+00  1310s
  31   1.55245984e+08  1.49819461e+08  3.91e-03 5.82e-11  1.36e+00  1348s
```

32	1.54470836e+08	1.50002253e+08	3.11e-03	5.82e-11	1.12e+00	1387s
33	1.53679063e+08	1.50381388e+08	2.31e-03	5.82e-11	8.28e-01	1429s
34	1.53095884e+08	1.50644190e+08	1.72e-03	5.82e-11	6.16e-01	1471s
35	1.52708811e+08	1.50840520e+08	1.33e-03	5.82e-11	4.69e-01	1519s
36	1.52428302e+08	1.50909302e+08	1.05e-03	5.82e-11	3.82e-01	1562s
37	1.52242587e+08	1.50970785e+08	8.69e-04	5.82e-11	3.19e-01	1605s
38	1.52062173e+08	1.51036387e+08	6.88e-04	7.28e-11	2.58e-01	1650s
39	1.51928679e+08	1.51095511e+08	5.56e-04	5.82e-11	2.09e-01	1695s
40	1.51838045e+08	1.51130875e+08	4.65e-04	5.82e-11	1.78e-01	1739s
41	1.51735798e+08	1.51168634e+08	3.65e-04	5.82e-11	1.42e-01	1783s
42	1.51673153e+08	1.51202357e+08	3.03e-04	5.82e-11	1.18e-01	1826s
43	1.51608964e+08	1.51233764e+08	2.41e-04	5.82e-11	9.42e-02	1872s
44	1.51575935e+08	1.51253510e+08	2.09e-04	5.82e-11	8.10e-02	1914s
45	1.51542481e+08	1.51266813e+08	1.77e-04	7.28e-11	6.92e-02	1956s
46	1.51514517e+08	1.51278544e+08	1.50e-04	8.37e-11	5.93e-02	2001s
47	1.51483685e+08	1.51290802e+08	1.20e-04	6.18e-11	4.84e-02	2044s
48	1.51451929e+08	1.51300418e+08	8.93e-05	6.15e-11	3.81e-02	2090s
49	1.51438711e+08	1.51312144e+08	7.63e-05	1.76e-10	3.18e-02	2133s
50	1.51426307e+08	1.51320949e+08	6.44e-05	2.62e-10	2.65e-02	2176s
51	1.51415671e+08	1.51327921e+08	5.44e-05	4.85e-10	2.20e-02	2220s
52	1.51399739e+08	1.51333182e+08	3.92e-05	6.39e-10	1.67e-02	2265s
53	1.51393823e+08	1.51339716e+08	3.35e-05	1.09e-09	1.36e-02	2311s
54	1.51381219e+08	1.51343131e+08	2.15e-05	1.52e-09	9.57e-03	2356s
55	1.51375694e+08	1.51346920e+08	1.63e-05	2.06e-09	7.23e-03	2405s
56	1.51371111e+08	1.51349155e+08	1.19e-05	2.91e-09	5.51e-03	2451s
57	1.51368698e+08	1.51350265e+08	9.64e-06	3.44e-09	4.63e-03	2495s
58	1.51368501e+08	1.51351293e+08	9.46e-06	3.97e-09	4.32e-03	2537s
59	1.51366922e+08	1.51352265e+08	7.96e-06	4.76e-09	3.68e-03	2580s
60	1.51365949e+08	1.51353312e+08	7.03e-06	5.86e-09	3.17e-03	2624s
61	1.51365098e+08	1.51353573e+08	6.22e-06	6.19e-09	2.89e-03	2665s
62	1.51363194e+08	1.51354469e+08	4.41e-06	7.28e-09	2.19e-03	2710s
63	1.51362371e+08	1.51354960e+08	3.63e-06	7.54e-09	1.86e-03	2752s
64	1.51361717e+08	1.51355602e+08	3.01e-06	7.50e-09	1.54e-03	2795s
65	1.51360611e+08	1.51356397e+08	1.99e-06	7.02e-09	1.06e-03	2843s
66	1.51360053e+08	1.51357029e+08	1.47e-06	5.93e-09	7.60e-04	2889s
67	1.51359657e+08	1.51357225e+08	1.10e-06	5.44e-09	6.11e-04	2932s
68	1.51359375e+08	1.51357544e+08	8.28e-07	4.47e-09	4.60e-04	2974s
69	1.51359061e+08	1.51357848e+08	5.38e-07	1.99e-08	3.05e-04	3019s
70	1.51358966e+08	1.51357990e+08	1.46e-06	3.59e-08	2.45e-04	3064s
71	1.51358816e+08	1.51358057e+08	1.63e-06	3.26e-08	1.91e-04	3111s
72	1.51358752e+08	1.51358140e+08	1.42e-06	2.75e-08	1.54e-04	3153s
73	1.51358550e+08	1.51358387e+08	1.75e-06	1.31e-08	4.08e-05	3206s
74	1.51358515e+08	1.51358436e+08	8.40e-07	4.94e-09	2.00e-05	3251s
75	1.51358485e+08	1.51358462e+08	9.50e-08	2.17e-09	5.82e-06	3295s
76	1.51358482e+08	1.51358481e+08	1.26e-07	1.66e-09	2.58e-07	3348s
77	1.51358482e+08	1.51358482e+08	2.25e-06	1.70e-09	1.11e-08	3396s
78	1.51358482e+08	1.51358482e+08	2.87e-08	1.73e-09	3.86e-11	3438s

```
Barrier solved model in 78 iterations and 3438.13 seconds
Optimal objective 1.51358482e+08


Root crossover log...

   13709 DPushes remaining with DInf 0.0000000e+00              3439s
       0 DPushes remaining with DInf 3.0967519-12               3440s

    1351 PPushes remaining with PInf 3.0772775e-04              3440s
       0 PPushes remaining with PInf 0.0000000e+00              3440s

  Push phase complete: Pinf 0.0000000e+00, Dinf 8.3569737e-04   3440s


Root simplex log...

Iteration    Objective       Primal Inf.    Dual Inf.      Time
   15008    1.5135848e+08   0.000000e+00   8.356974e-04   3440s
   15013    1.5135848e+08   0.000000e+00   0.000000e+00   3440s
   15013    1.5135848e+08   0.000000e+00   0.000000e+00   3441s

Root relaxation: objective 1.513585e+08, 15013 iterations, 3428.97 seconds

    Nodes    |    Current Node    |     Objective Bounds      |     Work
 Expl Unexpl |  Obj  Depth IntInf | Incumbent    BestBd   Gap | It/Node Time

     0     0 1.5136e+08    0   25 1.8521e+08 1.5136e+08  18.3%     - 3443s
H    0     0                    1.543105e+08 1.5136e+08  1.91%     [illegible]

Explored 0 nodes (15013 simplex iterations) in 3462.25 seconds
Thread count was 4 (of 4 available processors)

Solution count 2: 1.5431e+08 1.85209e+08
Pool objective bound 1.51358e+08

Optimal solution found (tolerance 7.00e-02)
Best objective 1.543104731084e+08, best bound 1.513584812246e+08, gap 1.9130%

Wrote result file 'MediumAdjSol.sol'
```

A.6.3 Large-Scale Case Study

```
Gurobi 7.0.0 (win64) logging started 07/26/17 07:56:54
```

```
Set parameter MIPGAP to value 0.009
Set parameter Method to value 2

Gurobi Optimizer version 7.0.0 build v7.0.0rc4 (win64)
Copyright (c) 2016, Gurobi Optimization, Inc.

Read LP format model from file Large.lp
Reading time = 34.24 seconds
P1: 1009080 rows, 4885920 columns, 14390831 nonzeros
Optimize a model with 1009080 rows, 4885920 columns and 14390831 nonzeros
Variable types: 4885650 continuous, 270 integer (0 binary)
Coefficient statistics:
  Matrix range     [3e-03, 8e+05]
  Objective range  [8e-04, 3e+05]
  Bounds range     [1e+00, 1e+00]
  RHS range        [2e-04, 7e+06]
Found heuristic solution: objective 5.17807e+08
Presolve removed 18458 rows and 494579 columns (presolve time = 8s) ...
Presolve removed 18458 rows and 494579 columns (presolve time = 11s) ...
Presolve removed 18459 rows and 494579 columns (presolve time = 17s) ...
Presolve removed 18459 rows and 494579 columns (presolve time = 20s) ...
Presolve removed 18459 rows and 494579 columns (presolve time = 28s) ...
Presolve removed 18459 rows and 494579 columns
Presolve time: 27.67s
Presolved: 990621 rows, 4391341 columns, 13013901 nonzeros
Variable types: 4391071 continuous, 270 integer (270 binary)
Presolve removed 0 rows and 0 columns (presolve time = 6s) ...
Presolved: 990621 rows, 4391341 columns, 13013901 nonzeros

Root barrier log...

Elapsed ordering time = 5s
Elapsed ordering time = 31s
Elapsed ordering time = 36s
Elapsed ordering time = 40s
Elapsed ordering time = 45s
Elapsed ordering time = 65s
Elapsed ordering time = 71s
Elapsed ordering time = 75s
Elapsed ordering time = 80s
Elapsed ordering time = 85s
Elapsed ordering time = 102s
Elapsed ordering time = 105s
Elapsed ordering time = 110s
Elapsed ordering time = 115s
Elapsed ordering time = 120s
```

```
Elapsed ordering time = 137s
Elapsed ordering time = 143s
Elapsed ordering time = 145s
Elapsed ordering time = 150s
Elapsed ordering time = 155s
Elapsed ordering time = 171s
Elapsed ordering time = 179s
Elapsed ordering time = 181s
Elapsed ordering time = 185s
Elapsed ordering time = 190s
Ordering time: 192.99s

Barrier statistics:
 AA' NZ     : 1.384e+07
 Factor NZ  : 1.028e+09 (roughly 10.0 GBytes of memory)
 Factor Ops : 2.512e+13 (roughly 400 seconds per iteration)
 Threads    : 2

                  Objective                  Residual
Iter       Primal          Dual          Primal    Dual     Compl     Time
   0   5.25469827e+15 -1.67765088e+17  3.96e+08 4.29e+01  5.91e+10  1106s
   1   3.55581219e+14 -9.19605577e+14  2.51e+07 5.82e-11  2.03e+09  2258s
   2   2.90509441e+13 -1.43012111e+14  1.54e+06 1.16e-10  1.32e+08  3425s
   3   5.20914468e+12 -5.16786731e+12  5.62e+04 5.82e-11  5.42e+06  5321s
   4   1.69062462e+11 -2.55932982e+12  7.26e+02 5.82e-11  3.34e+05  7354s
   5   5.48394365e+10 -3.79667132e+11  2.06e+01 1.16e-10  4.50e+04  10196s
   6   2.93571852e+10 -1.24734029e+11  1.03e+01 5.82e-11  1.59e+04  11169s
   7   1.01478532e+10 -6.94267138e+10  2.40e+00 1.16e-10  8.19e+03  13338s
   8   8.32954830e+09 -2.64895920e+10  1.73e+00 5.82e-11  3.58e+03  14522s
   9   6.47914217e+09 -1.20737125e+10  1.08e+00 1.16e-10  1.91e+03  15747s
  10   5.22482967e+09 -3.93701754e+09  7.33e-01 1.16e-10  9.43e+02  17799s
  11   3.60386765e+09 -1.21221356e+09  3.84e-01 1.16e-10  4.95e+02  19593s
  12   2.72258566e+09 -3.95625745e+08  2.81e-01 5.82e-11  3.21e+02  21895s
  13   2.25826413e+09 -1.94439254e+08  2.18e-01 5.82e-11  2.52e+02  23110s
  14   1.48826123e+09  1.34126429e+08  1.18e-01 5.82e-11  1.39e+02  26024s
  15   1.26435511e+09  2.14553673e+08  9.32e-02 5.82e-11  1.08e+02  27710s
  16   1.04545832e+09  2.77280497e+08  6.93e-02 5.82e-11  7.90e+01  29740s
  17   9.75514464e+08  2.97197471e+08  6.16e-02 5.82e-11  6.98e+01  31867s
  18   7.93676951e+08  3.43619587e+08  4.18e-02 5.82e-11  4.63e+01  33839s
  19   6.75526988e+08  3.61755550e+08  2.91e-02 5.82e-11  3.23e+01  35837s
  20   6.07278656e+08  3.80586601e+08  2.13e-02 5.82e-11  2.33e+01  37858s
  21   5.19984184e+08  3.96529049e+08  1.06e-02 5.82e-11  1.27e+01  39841s
  22   4.98313609e+08  4.01971884e+08  8.30e-03 1.16e-10  9.91e+00  41597s
  23   4.77328160e+08  4.13154667e+08  5.71e-03 5.82e-11  6.60e+00  43747s
  24   4.58677438e+08  4.17736435e+08  3.49e-03 5.82e-11  4.21e+00  45468s
  25   4.51625652e+08  4.20023112e+08  2.68e-03 2.60e-08  3.25e+00  47459s
  26   4.45059339e+08  4.22800283e+08  1.91e-03 2.78e-08  2.29e+00  49645s
```

27	4.42284902e+08	4.24085677e+08	1.59e-03	3.44e-08	1.87e+00	51352s
28	4.39787814e+08	4.24786802e+08	1.31e-03	3.75e-08	1.54e+00	52942s
29	4.36967653e+08	4.25442570e+08	9.83e-04	3.77e-08	1.19e+00	54664s
30	4.36499506e+08	4.25867723e+08	9.29e-04	4.06e-08	1.09e+00	55997s
31	4.34827533e+08	4.26178623e+08	7.43e-04	4.94e-08	8.90e-01	57714s
32	4.33005245e+08	4.26607690e+08	5.43e-04	5.20e-08	6.58e-01	60017s
33	4.32462730e+08	4.26835147e+08	4.84e-04	4.84e-08	5.79e-01	61481s
34	4.31681722e+08	4.27059312e+08	3.99e-04	4.30e-08	4.76e-01	63474s
35	4.31324767e+08	4.27186359e+08	3.61e-04	4.09e-08	4.26e-01	64956s
36	4.30943994e+08	4.27246915e+08	3.21e-04	4.18e-08	3.80e-01	66427s
37	4.30438976e+08	4.27337595e+08	2.67e-04	4.25e-08	3.19e-01	67759s
38	4.30033351e+08	4.27438806e+08	2.23e-04	4.89e-08	2.67e-01	69486s
39	4.29724744e+08	4.27515366e+08	1.90e-04	5.04e-08	2.27e-01	71152s
40	4.29583738e+08	4.27603867e+08	1.74e-04	4.81e-08	2.04e-01	72665s
41	4.29302531e+08	4.27654169e+08	1.44e-04	4.44e-08	1.70e-01	74845s
42	4.29076158e+08	4.27689321e+08	1.20e-04	3.68e-08	1.43e-01	76658s
43	4.28927692e+08	4.27724945e+08	1.03e-04	3.09e-08	1.24e-01	78125s
44	4.28834125e+08	4.27750723e+08	9.34e-05	2.34e-08	1.11e-01	79474s
45	4.28712779e+08	4.27794367e+08	8.00e-05	1.76e-08	9.45e-02	81590s
46	4.28596641e+08	4.27822311e+08	6.74e-05	1.35e-08	7.97e-02	83326s
47	4.28478980e+08	4.27837927e+08	5.50e-05	1.32e-08	6.59e-02	84930s
48	4.28417394e+08	4.27848805e+08	4.83e-05	1.28e-08	5.85e-02	86388s
49	4.28364061e+08	4.27868143e+08	4.29e-05	1.20e-08	5.10e-02	87972s
50	4.28309795e+08	4.27880823e+08	3.70e-05	1.18e-08	4.41e-02	89465s
51	4.28257079e+08	4.27894266e+08	3.13e-05	1.18e-08	3.73e-02	91083s
52	4.28212458e+08	4.27902305e+08	2.64e-05	1.18e-08	3.19e-02	92300s
53	4.28185985e+08	4.27916158e+08	2.35e-05	1.14e-08	2.78e-02	93879s
54	4.28153480e+08	4.27923283e+08	1.96e-05	1.10e-08	2.37e-02	95533s
55	4.28130395e+08	4.27929186e+08	1.70e-05	1.06e-08	2.07e-02	97011s
56	4.28099130e+08	4.27937711e+08	1.35e-05	9.89e-09	1.66e-02	99048s
57	4.28080010e+08	4.27940903e+08	1.20e-05	9.58e-09	1.43e-02	100395s
58	4.28069835e+08	4.27945095e+08	1.66e-05	9.14e-09	1.28e-02	101792s
59	4.28057021e+08	4.27949977e+08	1.43e-05	8.55e-09	1.10e-02	103857s
60	4.28045785e+08	4.27954466e+08	1.51e-05	8.13e-09	9.39e-03	105686s
61	4.28037468e+08	4.27957696e+08	1.64e-05	7.81e-09	8.21e-03	107065s
62	4.28029174e+08	4.27958885e+08	2.45e-05	7.70e-09	7.23e-03	108395s
63	4.28018658e+08	4.27961833e+08	1.40e-05	7.39e-09	5.85e-03	110056s
64	4.28011652e+08	4.27964166e+08	1.79e-05	7.31e-09	4.88e-03	111440s
65	4.28006851e+08	4.27965382e+08	1.68e-05	7.22e-09	4.27e-03	112770s
66	4.28000363e+08	4.27968997e+08	1.06e-05	9.08e-09	3.23e-03	114506s
67	4.27994862e+08	4.27970654e+08	8.47e-06	1.03e-08	2.49e-03	116407s
68	4.27990113e+08	4.27972558e+08	2.43e-06	1.02e-08	1.81e-03	118390s
69	4.27988088e+08	4.27973574e+08	2.87e-06	9.41e-09	1.49e-03	120142s
70	4.27985936e+08	4.27974437e+08	2.77e-06	8.42e-09	1.18e-03	122922s
71	4.27984745e+08	4.27974747e+08	2.43e-06	7.86e-09	1.03e-03	124258s
72	4.27983200e+08	4.27975292e+08	2.00e-06	6.83e-09	8.13e-04	126631s
73	4.27982229e+08	4.27975760e+08	1.49e-06	6.40e-09	6.65e-04	128518s

74	4.27982133e+08	4.27975981e+08	1.48e-06	6.23e-09	6.33e-04	129972s
75	4.27981216e+08	4.27976152e+08	1.41e-06	6.13e-09	5.21e-04	131698s
76	4.27980914e+08	4.27976296e+08	1.52e-06	5.88e-09	4.75e-04	133028s
77	4.27980430e+08	4.27976476e+08	9.40e-07	5.58e-09	4.07e-04	134514s
78	4.27980194e+08	4.27976666e+08	7.63e-07	5.96e-09	3.63e-04	135858s
79	4.27979650e+08	4.27976814e+08	4.84e-07	7.52e-09	2.92e-04	137336s
80	4.27979310e+08	4.27977039e+08	7.92e-07	1.23e-08	2.34e-04	139317s
81	4.27979064e+08	4.27977163e+08	8.81e-07	1.46e-08	1.96e-04	140752s
82	4.27979008e+08	4.27977186e+08	8.68e-07	1.50e-08	1.87e-04	142092s
83	4.27978808e+08	4.27977261e+08	8.31e-07	1.62e-08	1.59e-04	143806s
84	4.27978684e+08	4.27977302e+08	7.56e-07	1.65e-08	1.42e-04	145524s
85	4.27978542e+08	4.27977358e+08	7.56e-07	1.68e-08	1.22e-04	146980s
86	4.27978428e+08	4.27977441e+08	6.53e-07	1.74e-08	1.02e-04	148841s
87	4.27978292e+08	4.27977505e+08	5.26e-07	1.85e-08	8.10e-05	150437s
88	4.27978143e+08	4.27977547e+08	8.22e-07	1.92e-08	6.13e-05	152818s
89	4.27978100e+08	4.27977571e+08	8.26e-07	1.89e-08	5.45e-05	154144s
90	4.27978045e+08	4.27977584e+08	7.79e-07	1.86e-08	4.74e-05	155474s
91	4.27978038e+08	4.27977592e+08	7.72e-07	1.83e-08	4.58e-05	157191s
92	4.27977978e+08	4.27977605e+08	7.50e-07	1.77e-08	3.83e-05	158653s
93	4.27977955e+08	4.27977614e+08	7.11e-07	1.70e-08	3.50e-05	159853s
94	4.27977917e+08	4.27977631e+08	8.59e-07	1.59e-08	2.94e-05	161469s
95	4.27977873e+08	4.27977648e+08	6.17e-07	1.37e-08	2.32e-05	163619s
96	4.27977839e+08	4.27977659e+08	5.03e-07	1.14e-08	1.86e-05	165643s
97	4.27977813e+08	4.27977670e+08	4.05e-07	9.31e-09	1.47e-05	167764s
98	4.27977793e+08	4.27977676e+08	3.24e-07	8.24e-09	1.20e-05	170490s
99	4.27977781e+08	4.27977681e+08	2.77e-07	7.14e-09	1.03e-05	172552s
100	4.27977774e+08	4.27977684e+08	2.45e-07	6.60e-09	9.26e-06	174242s
101	4.27977764e+08	4.27977690e+08	2.05e-07	5.27e-09	7.59e-06	178573s
102	4.27977755e+08	4.27977694e+08	1.71e-07	4.47e-09	6.36e-06	180276s
103	4.27977751e+08	4.27977696e+08	1.52e-07	3.98e-09	5.66e-06	182123s
104	4.27977741e+08	4.27977699e+08	1.10e-07	3.26e-09	4.27e-06	185981s
105	4.27977737e+08	4.27977700e+08	9.54e-08	3.03e-09	3.78e-06	187739s
106	4.27977734e+08	4.27977703e+08	8.27e-08	2.61e-09	3.24e-06	191638s
107	4.27977731e+08	4.27977704e+08	7.09e-08	2.31e-09	2.80e-06	193954s
108	4.27977728e+08	4.27977704e+08	5.90e-08	2.25e-09	2.47e-06	196613s
109	4.27977724e+08	4.27977706e+08	4.13e-08	1.83e-09	1.81e-06	198589s
110	4.27977722e+08	4.27977708e+08	3.39e-08	1.51e-09	1.47e-06	200322s
111	4.27977721e+08	4.27977708e+08	2.98e-08	1.43e-09	1.33e-06	202246s
112	4.27977720e+08	4.27977709e+08	2.52e-08	1.25e-09	1.14e-06	204062s
113	4.27977719e+08	4.27977709e+08	1.87e-08	1.11e-09	9.75e-07	206498s
114	4.27977718e+08	4.27977710e+08	1.62e-08	1.04e-09	8.80e-07	208717s
115	4.27977717e+08	4.27977710e+08	1.36e-08	8.51e-10	7.20e-07	210859s
116	4.27977717e+08	4.27977711e+08	1.14e-08	8.01e-10	6.33e-07	213281s
117	4.27977717e+08	4.27977711e+08	1.12e-08	7.54e-10	6.07e-07	215535s
118	4.27977716e+08	4.27977711e+08	9.35e-09	6.89e-10	5.25e-07	218465s
119	4.27977716e+08	4.27977712e+08	7.97e-09	5.80e-10	4.38e-07	220898s
120	4.27977715e+08	4.27977712e+08	4.01e-09	4.21e-10	2.58e-07	227060s

```
 121   4.27977714e+08  4.27977712e+08  3.21e-09 3.49e-10  2.07e-07  231331s
 122   4.27977714e+08  4.27977713e+08  7.48e-07 1.89e-10  1.13e-07  233536s

Barrier solved model in 122 iterations and 233537.63 seconds
Optimal objective 4.27977714e+08


Root crossover log...

   40054 DPushes remaining with DInf 0.0000000e+00              233540s
   10263 DPushes remaining with DInf 0.0000000e+00              233545s
       0 DPushes remaining with DInf 4.1169023e-11              233549s

 1577882 PPushes remaining with PInf 8.2022369e-02              233550s
  328300 PPushes remaining with PInf 4.6782478e-01              233552s
  100266 PPushes remaining with PInf 4.8792527e-01              233555s
       0 PPushes remaining with PInf 2.1316700e-04              233560s

  Push phase complete: Pinf 2.1316700e-04, Dinf 3.8982096e+03 233560s


Root simplex log...

Iteration    Objective       Primal Inf.    Dual Inf.      Time
 1617881    4.2797771e+08   0.000000e+00   3.898210e+03 233560s
 1619935    4.2797771e+08   0.000000e+00   0.000000e+00 233563s
 1619935    4.2797771e+08   0.000000e+00   0.000000e+00 233568s

Root relaxation: objective 4.279777e+08, 1619935 iterations, 233537.31 seconds
Total elapsed time = 233570.81s

    Nodes    |    Current Node    |     Objective Bounds      |     Work
 Expl Unexpl |  Obj  Depth IntInf | Incumbent    BestBd   Gap | It/Node Time

     0     0 4.2798e+08    0   26 5.1781e+08 4.2798e+08  17.3%     - 233575s
H    0     0                    4.298386e+08 4.2798e+08  0.43%     - 233632s

Explored 0 nodes (1619935 simplex iterations) in 233637.25 seconds
Thread count was 4 (of 4 available processors)

Solution count 2: 4.29839e+08 5.17807e+08
Pool objective bound 4.27978e+08

Optimal solution found (tolerance 9.00e-03)
Best objective 4.298386022175e+08, best bound 4.279777130486e+08, gap 0.4329%

Wrote result file 'LargeLaptopNew.sol'
```

A.6.4 Katrina Case Study

```
Gurobi 7.0.0 (win64) logging started 07/24/17 14:22:23

Set parameter Method to value 2
Set parameter MIPGap to value 0.007

Gurobi Optimizer version 7.0.0 build v7.0.0rc4 (win64)
Copyright (c) 2016, Gurobi Optimization, Inc.

Read LP format model from file ShortTerm.lp
Reading time = 36.93 seconds
P1: 1009080 rows, 4885920 columns, 14390895 nonzeros
Optimize a model with 1009080 rows, 4885920 columns and 14390895 nonzeros
Variable types: 4885650 continuous, 270 integer (0 binary)
Coefficient statistics:
  Matrix range     [9e-02, 8e+05]
  Objective range  [3e-04, 3e+05]
  Bounds range     [1e+00, 1e+00]
  RHS range        [5e-02, 4e+06]
Found heuristic solution: objective 4.28383e+08
Presolve removed 17723 rows and 88030 columns (presolve time = 5s) ...
Presolve removed 17723 rows and 492468 columns (presolve time = 10s) ...
Presolve removed 17723 rows and 492468 columns (presolve time = 15s) ...
Presolve removed 17723 rows and 492468 columns (presolve time = 22s) ...
Presolve removed 17723 rows and 492468 columns (presolve time = 25s) ...
Presolve removed 17723 rows and 492468 columns (presolve time = 33s) ...
Presolve removed 17723 rows and 492468 columns
Presolve time: 32.88s
Presolved: 991357 rows, 4393452 columns, 13020471 nonzeros
Variable types: 4393182 continuous, 270 integer (270 binary)
Presolve removed 0 rows and 0 columns (presolve time = 6s) ...
Presolved: 991357 rows, 4393452 columns, 13020471 nonzeros

Root barrier log...

Elapsed ordering time = 5s
Elapsed ordering time = 32s
Elapsed ordering time = 37s
Elapsed ordering time = 40s
Elapsed ordering time = 45s
Elapsed ordering time = 68s
Elapsed ordering time = 71s
Elapsed ordering time = 75s
```

```
Elapsed ordering time = 80s
Elapsed ordering time = 85s
Elapsed ordering time = 103s
Elapsed ordering time = 106s
Elapsed ordering time = 111s
Elapsed ordering time = 115s
Elapsed ordering time = 120s
Elapsed ordering time = 138s
Elapsed ordering time = 144s
Elapsed ordering time = 145s
Elapsed ordering time = 150s
Elapsed ordering time = 155s
Elapsed ordering time = 172s
Elapsed ordering time = 176s
Elapsed ordering time = 181s
Elapsed ordering time = 185s
Elapsed ordering time = 190s
Ordering time: 191.85s

Barrier statistics:
 AA' NZ     : 1.385e+07
 Factor NZ  : 1.020e+09 (roughly 10.0 GBytes of memory)
 Factor Ops : 2.463e+13 (roughly 400 seconds per iteration)
 Threads    : 2

                  Objective                Residual
Iter       Primal          Dual         Primal    Dual     Compl     Time
   0   5.30471060e+17 -5.19675427e+16  3.00e+10 1.14e-13  3.24e+12  1058s
   1   2.14107304e+16 -3.28633936e+16  1.16e+09 5.82e-11  8.47e+10  2597s
   2   4.89394623e+15 -5.20326648e+15  2.64e+08 1.16e-10  1.89e+10  3879s
   3   6.63456911e+14 -6.70520291e+14  3.57e+07 1.17e-09  2.64e+09  5153s
   4   3.04492295e+13 -1.26835599e+14  1.55e+06 5.82e-11  1.37e+08  7016s
   5   1.54647524e+13 -1.49516644e+13  7.34e+05 1.16e-10  5.65e+07  8526s
   6   9.85468897e+11 -2.92068520e+12  4.26e+03 3.49e-10  6.60e+05  10018s
   7   6.64923203e+10 -9.84894775e+11  1.94e+02 1.16e-10  1.15e+05  12198s
   8   2.52084597e+10 -2.34766985e+11  1.26e+01 1.16e-10  2.69e+04  14598s
   9   1.02571845e+10 -7.48125688e+10  1.09e+00 5.82e-11  8.75e+03  16590s
  10   8.73360682e+09 -4.65233188e+10  5.41e-01 1.16e-10  5.68e+03  17875s
  11   4.14362200e+09 -2.08507503e+09  4.45e-02 1.16e-10  6.40e+02  20332s
  12   3.66114983e+09 -1.29473113e+09  3.58e-02 5.82e-11  5.10e+02  21494s
  13   3.52599631e+09 -1.07621627e+09  3.37e-02 5.82e-11  4.73e+02  22634s
  14   1.80055587e+09 -2.56500682e+08  1.57e-02 5.82e-11  2.12e+02  24312s
  15   6.87671143e+08  3.53160529e+08  1.94e-03 5.82e-11  3.44e+01  26710s
  16   5.87092051e+08  3.57555673e+08  1.31e-03 5.82e-11  2.36e+01  27969s
  17   4.98484469e+08  3.66851051e+08  7.40e-04 5.82e-11  1.35e+01  29458s
  18   4.68695574e+08  3.70096966e+08  5.48e-04 5.82e-11  1.01e+01  30748s
  19   4.41736301e+08  3.74942611e+08  3.95e-04 5.82e-11  6.87e+00  32276s
```

20	4.20571158e+08	3.78129420e+08	2.50e-04	7.28e-11	4.36e+00	33547s
21	4.06232288e+08	3.79988650e+08	1.63e-04	6.73e-11	2.70e+00	35787s
22	4.00851043e+08	3.80930510e+08	1.27e-04	6.91e-11	2.05e+00	37309s
23	3.94984075e+08	3.81733024e+08	8.98e-05	8.00e-11	1.36e+00	38881s
24	3.91064145e+08	3.82155884e+08	6.07e-05	7.21e-11	9.16e-01	40528s
25	3.88022351e+08	3.82497479e+08	3.84e-05	7.28e-11	5.68e-01	42604s
26	3.86900355e+08	3.82619070e+08	3.03e-05	5.82e-11	4.40e-01	44005s
27	3.85963886e+08	3.82679426e+08	2.35e-05	7.64e-11	3.38e-01	45547s
28	3.85320529e+08	3.82742077e+08	1.85e-05	6.55e-11	2.65e-01	46796s
29	3.84759165e+08	3.82800702e+08	1.45e-05	8.37e-11	2.01e-01	48335s
30	3.84313434e+08	3.82834570e+08	1.10e-05	7.55e-11	1.52e-01	49876s
31	3.83879300e+08	3.82860262e+08	7.69e-06	8.37e-11	1.05e-01	51697s
32	3.83697287e+08	3.82877978e+08	6.50e-06	6.73e-11	8.42e-02	53027s
33	3.83510131e+08	3.82892287e+08	4.73e-06	7.82e-11	6.35e-02	54556s
34	3.83417100e+08	3.82900995e+08	4.15e-06	6.91e-11	5.31e-02	55894s
35	3.83354286e+08	3.82918239e+08	3.89e-06	6.55e-11	4.48e-02	57242s
36	3.83269165e+08	3.82924839e+08	3.55e-06	6.55e-11	3.54e-02	58872s
37	3.83239555e+08	3.82930326e+08	3.51e-06	8.91e-11	3.18e-02	60225s
38	3.83205525e+08	3.82938220e+08	3.15e-06	6.37e-11	2.75e-02	61878s
39	3.83172202e+08	3.82944455e+08	2.99e-06	7.09e-11	2.34e-02	63191s
40	3.83144166e+08	3.82945682e+08	2.83e-06	7.09e-11	2.04e-02	64419s
41	3.83116464e+08	3.82950723e+08	2.97e-06	9.19e-11	1.70e-02	65671s
42	3.83089414e+08	3.82953536e+08	2.86e-06	7.82e-11	1.40e-02	67075s
43	3.83072780e+08	3.82956456e+08	2.74e-06	9.28e-11	1.20e-02	68447s
44	3.83054657e+08	3.82958647e+08	2.27e-06	6.18e-11	9.87e-03	69763s
45	3.83034791e+08	3.82960479e+08	1.68e-06	8.91e-11	7.64e-03	71420s
46	3.83027661e+08	3.82961215e+08	1.90e-06	8.37e-11	6.83e-03	72787s
47	3.83017042e+08	3.82962666e+08	2.14e-06	8.91e-11	5.59e-03	74197s
48	3.83007684e+08	3.82963102e+08	1.90e-06	9.09e-11	4.58e-03	75417s
49	3.82992275e+08	3.82963877e+08	1.54e-06	2.45e-10	2.92e-03	77203s
50	3.82985570e+08	3.82964758e+08	1.55e-06	5.38e-10	2.14e-03	78867s
51	3.82981177e+08	3.82965098e+08	1.57e-06	7.31e-10	1.65e-03	80710s
52	3.82977409e+08	3.82965542e+08	2.32e-06	8.09e-10	1.22e-03	82401s
53	3.82974908e+08	3.82965744e+08	3.09e-06	8.28e-10	9.42e-04	83677s
54	3.82973408e+08	3.82965910e+08	3.74e-06	9.01e-10	7.71e-04	85109s
55	3.82972445e+08	3.82966153e+08	3.60e-06	9.57e-10	6.47e-04	86332s
56	3.82969934e+08	3.82966408e+08	2.45e-06	1.22e-09	3.62e-04	87924s
57	3.82969196e+08	3.82966484e+08	1.94e-06	1.28e-09	2.79e-04	89490s
58	3.82967916e+08	3.82966535e+08	1.12e-06	1.20e-09	1.42e-04	91345s
59	3.82967493e+08	3.82966575e+08	1.06e-06	1.22e-09	9.43e-05	92914s
60	3.82967081e+08	3.82966595e+08	6.36e-07	1.12e-09	4.99e-05	94979s
61	3.82966784e+08	3.82966614e+08	1.69e-07	1.45e-09	1.75e-05	96702s
62	3.82966639e+08	3.82966630e+08	1.89e-07	6.54e-10	9.70e-07	99685s
63	3.82966632e+08	3.82966632e+08	9.66e-07	8.73e-11	3.72e-08	103639s
64	3.82966632e+08	3.82966632e+08	1.58e-06	7.28e-11	7.21e-09	105479s

```
Barrier solved model in 64 iterations and 105480.74 seconds
```

```
Optimal objective 3.82966632e+08


Root crossover log...

   45718 DPushes remaining with DInf 0.0000000e+00            105484s
   19216 DPushes remaining with DInf 0.0000000e+00            105485s
    7290 DPushes remaining with DInf 0.0000000e+00            105490s
       0 DPushes remaining with DInf 1.8287474e-11            105495s

  237336 PPushes remaining with PInf 5.1614022e-02            105495s
   13976 PPushes remaining with PInf 6.1438729e-02            105496s
       0 PPushes remaining with PInf 0.0000000e+00            105496s

  Push phase complete: Pinf 0.0000000e+00, Dinf 2.0727572e+00 105497s


Root simplex log...

Iteration    Objective       Primal Inf.    Dual Inf.      Time
  283056    3.8296663e+08   0.000000e+00   2.072757e+00 105497s
  283398    3.8296663e+08   0.000000e+00   0.000000e+00 105499s
  283398    3.8296663e+08   0.000000e+00   0.000000e+00 105505s

Root relaxation: objective 3.829666e+08, 283398 iterations, 105469.00 seconds
Total elapsed time = 105506.20s

    Nodes    |    Current Node    |     Objective Bounds      |     Work
 Expl Unexpl |  Obj  Depth IntInf | Incumbent    BestBd   Gap | It/Node Time

     0     0 3.8297e+08    0   23 4.2838e+08 3.8297e+08  10.6%     - 105514s
H    0     0                    3.872088e+08 3.8297e+08  1.10%     - 105525s
H    0     0                    3.841701e+08 3.8297e+08  0.31%     - 105646s

Explored 0 nodes (283398 simplex iterations) in 105651.19 seconds
Thread count was 4 (of 4 available processors)

Solution count 3: 3.8417e+08 3.87209e+08 4.28383e+08
Pool objective bound 3.82967e+08

Optimal solution found (tolerance 7.00e-03)
Best objective 3.841700714954e+08, best bound 3.829666318226e+08, gap 0.3133%

Wrote result file 'Short.sol'
```

MIX
Papier aus verantwortungsvollen Quellen
Paper from responsible sources
FSC® C105338

If you have any concerns about our products,
you can contact us on
ProductSafety@springernature.com

In case Publisher is established outside the EU,
the EU authorized representative is:
Springer Nature Customer Service Center GmbH
Europaplatz 3, 69115 Heidelberg, Germany

Printed by Libri Plureos GmbH
in Hamburg, Germany